Modern Newspaper Editing and Production

by the same author

Modern Newspaper Practice

Modern Newspaper Editing and Production

F. W. HODGSON

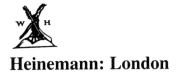

Heinemann: London

To Yvonne, for her unfailing support

William Heinemann Ltd
10 Upper Grosvenor Street, London W1X 9PA
LONDON   MELBOURNE   JOHANNESBURG   AUCKLAND

First published 1987
© F. W. Hodgson 1987

British Library Cataloguing in Publication Data
Hodgson, F. W.
  Modern newspaper editing and production.
  1. Journalism   2. Newspaper layout and
  typography   3. Newspapers
  I. Title
  070.4′1   PN4776

ISBN 0 434 90750 2

Typeset by Deltatype Ltd, Ellesmere Port, S. Wirral
Printed in Great Britain by
Redwood Burn Ltd, Trowbridge

CONTENTS

Preface   ix
Acknowledgements   x

1 Printing systems   1
Hot metal – The cold type revolution – The terminals – The computer – Photosetting – Web offset – Decentralized printing – Effect on editorial production

2 Choosing the materials   14
News sources – Staff reporters – Freelance reporters – Local correspondents – News agencies – Hand-outs – The newsroom – The copy-taster – Measuring the news – Patterns of news – Tasting methods – Electronic copy-tasting

3 Page planning and typography   33
Format – Who does what – Principles of design – Focal points – Order – Type character – Typography – Type use – Typesizes – Measure – Drawing the page

4 Pictures   56
Picture sources – Staff photographers – Freelance photographers – Picture agencies – Collected pictures – Hand-out pictures – Picture library – Photo briefing – Choosing pictures – Content – Composition – Balance – Tone – The design function – Uses of pictures – Pictures at work – Picture editing – Cropping – Scaling – Retouching – Colour work – Ethics and taste – Graphics

5 Subediting: basic techniques   77
Getting it right – Order and shape – The intro – The delayed drop – Story sequence – Quoted speech – Geography – Time – Background – Casting off – Electronic editing – Marking up

6 Subediting: language and its traps 100
The sentence – Participles – Pronouns – Tenses – Verbs – Split infinitives – Qualifiers – Nouns – Prepositions – Other faults – Length – Paragraphs – Punctuation – Comma – Full stop – Semicolon – Colon – Dashes and brackets – Ellipses – Exclamation mark – Apostrophe – Question mark – Hyphen – Stresses

7 Subediting: the words on the page 119
The right word – Technical words – Foreign words – Circumlocutions – Synonyms – Clichés – Vogue words – Misused words – House style – Spellings – American words – Anglicized words – Capital letters – Numbers – Cyphers – Abbreviations – Typographical style – Trade names

8 Headlines: content and approach 137
The words – Subject – Verb – Special words – Omission of words – Nouns as adjectives – Symbols – Word accuracy – Headline punctuation – Headline abbreviations – Numbers – Content – Taste – Vital facts – The personal touch – Time – Location – Composing a headline – The direct approach – The oblique approach – Headline thoughts – Label headlines – Split headlines – Turn heads – Things to avoid

9 Headlines: The typographical challenge 152
Shape – Arrangement – Typestyle – Character counts – Spacing – The creative moment – Alternative words – Word list

10 Subediting: further techniques 170
Subeditors and the law – Libel – Contempt of court – The Official Secrets Act – Absolute privilege – Qualified privilege – Rewrites – Angling – Multiple copy sources – Bad copy – Electronic aids – Revising and editionizing – Rejigs – Captions – Caption typography – Contents bills

11 The running story 187
Handling the text

12 Putting it all together 199
Composing area – Make-up – Pictures – Headlines and text – Stone subbing – Paste-up cutting – Spacing – Printing the paper

13 Voices and audiences 213
Special vocabularies – Sport – Financial – The soft sell – Facetiousness – Is there a tabloid style? – Journalese

14 Features: planning and design 226
Copy sources – Planning – Control – Page design – Typography – Pictures – The text – Guiding the reader – Features workshop

15 Features: language and projection 242
Accuracy – Language – Ghost writing – Reader participation – Features and the law – Projection – Features headlines – The emotive phrase – The whimsical phrase – The informative phrase – The decorative phrase – The confessional

Glossary 254
Index 273

PREFACE

The rapid advance of computerized printing systems and the electronic editing facilities they offer make this a vital time to reassess editorial practices in newspapers. Though what used to be called 'new technology' is now commonplace, not all systems are being utilized at present to the same extent. There have been difficult human problems to solve. What is clear is that 'hot metal' as a means of type generation is now a thing of the past in Britain and most parts of the world, and that screen editing and phototypesetting have become concepts that have to be grasped by journalists.

The aim of this book is to re-examine the editing and editorial production of newspapers in the light of these changes. It will stress both the continuity of methods that are still necessary to produce well informed, well edited and well designed newspapers, and the exciting challenge that the new systems present in these areas.

The continuity theme is an important one at a time when many newspapers have gone through traumatic experiences in their move into the computer age. The new systems, though they are worlds apart from the old mechanical hot metal printing, are not a signal for a wholesale change of editorial approach, or for the production of new style or different newspapers. The thing to remember is that the machine has not taken over the newspaper but is being utilized by editors and their staffs to produce newspapers that read and look the way they want them to read and look.

The bonus, if the systems are used correctly, is that this may be done to an even higher degree of professional polish and reader satisfaction, and that the process may be faster and cheaper, thus making more newspapers and a greater choice available to readers.

F. W. Hodgson

ACKNOWLEDGEMENTS

I have not included a general bibliography in this textbook but I would be failing in my respects to my predecessors in this field if I did not acknowledge my debt to them. Harold Evans's five-volume work, *Editing and Design* (Heinemann), produced in the early seventies, remains a repository of great wisdom and expertise on the craft of newspaper journalism as it stood at the moment of onset of 'new technology'. Likewise of its period, Leslie Sellers's little *Simple Subs' Book* (Pergamon Press 1968) has whetted the appetite of many a young subeditor and is still valid in its essentials. Keith Waterhouse's *Daily Mirror Style* (Mirror Books 1981) reminds us of the unique contribution to journalism made by the popular tabloids, while Ted Bottomley and Anthony Loftus's *Journalists' Guide to the Use of English* (Express and Star Publications 1971) has many years left as an introduction to newspaper English.

I found the following works useful in my chapters dealing with the knotty problems of text editing that confront subeditors: Burchfield, Robert, *The Spoken Word*, a BBC Guide (BBC 1981); Carey, G. V., *Mind the Stop* (Penguin Books 1976); *Fowler's English Usage*, 2nd ed., revised and edited by Sir Ernest Gowers (Oxford University Press 1982); Gowers, Sir Ernest, *The Complete Plain Words*, revised ed. (Penguin Books 1977); Partridge, Eric, *Usage and Abusage* (Penguin Books 1969).

For permission to reprint material, I am indebted to the editors of the *Brighton Argus*, the *Daily Telegraph*, *Financial Times*, *Liverpool Echo*, *The Sunday Times*, *The Guardian*, *The Sun*, *Western Mail*, *South London Press*, *Birmingham Post*, *London Evening Standard* and *Daily Mail*. For special help and facilities I am particularly grateful to Portsmouth and Sunderland Newspapers Ltd, News Group Newspapers Ltd, and London Post (Printers) Ltd, and to Brian Thomas for taking the pictures of new technology at work.

1 PRINTING SYSTEMS

Between the writing of news stories and features and the printing of a newspaper lies the area that is called editorial production. It is inhabited by journalists who, for want of a better word, have been termed the processors, as opposed to the gatherers, of newspaper material.

They consist of the editor, the deputy editor and assistant editors, the departmental heads such as the news editor, features editor, picture editor and art editor; the layout artists and, finally, the subeditors, who are the journalists who do the day-to-day editing of the paper's content. These are the people who make and carry out the decisions and go through the procedures, of which the end product is the newspaper as it appears on the news-stands or is delivered through the letter-box.

The actual printing is only the final stage of this process. Yet it is an important stage, for journalists, more than any other writers, are closely involved with printing because of the daily deadlines of production and the visual side of their work. It is necessary, therefore, before examining in detail the techniques used in modern editorial production, that we should look at what has been happening in printing in the last few decades.

During this period the computer has wrought a revolution that has changed an old craft into a modern high-tech system in which the only recognizable thing left is that it generates type. Starting in 1961, and gathering pace through the 1960s, a new system based on developments in electronics and computer science began to replace traditional hot metal printing technology in the United States. Instead of text for newspapers (called copy) being set in lines of metal type on line-casting machines worked by operators, it began to be turned into 'cold type', which was type produced photographically inside a machine called a phototypesetter.

The change could not have been more fundamental. Pages had always been made up from the lines of type assembled in columns and from pictures and advertisements engraved on metal plates. Now they were put together from the text and headlines delivered by the photosetter as photographic prints, which were pasted on to make-up sheets along with prints of the pictures and advertisements. Instead of the printing plate for the presses being cast from a matrix, or mould, taken from a made-up metal page, it was derived from a photographic transparency of the pasted-up page. The new type printing surface was also made of plastic instead of metal.

If at first it seemed a fragile substitute for a technology that had been based on heavy metal and clanking machinery since the days of Caxton and Gutenberg, 'cold type' was soon to prove its detractors wrong. Within two decades it had been taken up for newspaper printing not only in America but in Europe and many other parts of the world. Today the manufacture of line-casting machines has ceased, hot metal technology is in the terminal stages of decline, and cold, or photoset, type has become the basis of the systems by which modern newspapers are printed.

HOT METAL

Hot metal printing technology arose out of the sound theory that to make multiple copies of the same piece of written work you had to create a durable image of it in reverse which could be used repeatedly to transfer an inked impression of itself on to paper. A lead-based alloy was the best substance for this since not only was it hard but it could be melted down and used again for a different piece of work when its purpose was at an end.

The economic way to produce this reverse image was to design separate metal 'master' letters which could be selected and arranged into words and sentences to reproduce the piece of written work of which copies were needed. Then, so that the letters could be used again, a metal alloy cast of each line could be taken showing the words in reverse. It was by assembling these lines together that the page took shape. The final stage was the transfer of the page image, by means of a matrix, or mould, to the printing plate, which showed the whole page image in reverse. From this, the piece of work could be printed on paper by ink impression with the letters appearing in positive.

The concept of printing by this method was one of the great technological breakthroughs of history. Its basis remained unchanged for hundreds of years, although it improved in quality and speed through constant refinement. Elegant metal alphabets for a variety of uses were designed by the great type-founders; paper manufacture improved, and so did the balance of the metal alloy used for casting. Printing presses themselves underwent improvement leading from the early manually operated flatbed presses to the great rotary presses capable of printing newspapers at up to 60,000 copies an hour.

The need for mass production of newspapers was a factor in the improvements in hot metal technology. The invention of the linotype machine by Ottmar Mergenthaler in Baltimore in 1884 took the sweat out of hand-composing of small reading type and enabled newspaper stories to be set by a keyboard operator at up to five lines a minute. The machine contained matrices of all the characters of a given type, which were brought together to form lines which were automatically cast inside the machine by the injection of hot metal. The lines were then ejected into a tray until the entire sequence of lines for the story was made up. Another machine enabled headlines, which still had to be hand-composed on to a 'stick', to be cast by slotting the stick into a hot metal caster.

Improvements were made in the material, or flong, used for taking the moulds from the pages that had been made up from the lines of type. More efficient moulding presses produced sharper moulds by applying the flong to the made-up pages at high pressure. Better quality was achieved in the casting of printing plates from the page moulds for use on the rotary presses.

Finally, efficient high-speed etching machines took over from the old craftsmen engravers so that the making of blocks, or engravings of the pictures, artwork and advertisements could be done in batches and thus speeded up to avoid delays in page make-up.

There were a number of advantages for newspapers with hot metal technology. The setting of copy by the line, or slug, enabled alterations to be made easily at page make-up stage by the substitution of lines and paragraphs. Setting errors, called literals – which had been marked on proofs of stories by the proof readers – could be set and dropped in the stories while the lines of type lay on their galleys, awaiting transfer to the pages.

During the make-up stage, while the page was being put together inside its metal frame, or chase, adjustments could be made easily to the page design and to the spacing. The printer in charge of making up the page – the stone hand – could manipulate the type so that stories could be altered in length and position before the page was locked together ready for moulding.

The use of the term 'stone' is an interesting survival from the early days of printing. It denotes the metal bench on which pages are made up under the hot metal system, and which formerly had stone tops to take the heavy weight. Hence stone hand. Hence also the stone subeditor, or stone-sub, the production journalist who supervises the page make-up to see that it matches the editorial layout and the editor's requirements. The term stone-sub is still used in offices that have long switched over to cold type and paste-up.

Another term which has survived is the word forme which is given to the made-up page (or pair of pages in tabloid newspapers) when ready for moulding. The 'set forme' denotes the last forme to be made ready for printing, i.e. the moment when the paper can go to press.

Another advantage of the hot metal system was that it was a tough system.

Its machines were heavy and lasted well, and the method of printing by heavy relief metal plates on rotary presses – letterpress – could stand long print runs without plate or machine breakage.

Yet there were snags as well. Plant was bulky and took up large areas, especially the scores of line-casters. Floors had to be reinforced, and there were insoluble problems of noise and dirt due to the use of metal and ink at the typesetting and make-up stage. But most of all, hard bargaining by print unions had led to high wage costs in what remained a labour-intensive industry with a highly perishable product. To print in two centres a big newspaper had to duplicate the same costly, labour-intensive space-taking operation. The cost of production meant that the return on capital for newspaper companies was generally poor – a fact which contributed to the decline in the number of newspaper titles both in Britain and America during the decades following the 1940s.

This was the situation when the computer burst upon the printing scene in America in 1961.

THE COLD TYPE REVOLUTION

Though at first there were shortcomings in computer capacity, the use of cold type promised to solve at a stroke the problem of noise, dirt, space, low profits and high production costs, and refinements soon took place to adapt it more closely to newspaper requirements. Using a cathode-ray tube to project the setting from type 'masters' as photographic images, the new photosetters could do the work of scores of heavy slow line-casters both more accurately and at greater speeds.

Moreover, the page produced in this way, which was itself a piece of film, interfaced conveniently with the much cheaper method of web-offset printing, which used smooth plastic-coated printing plates on to which the page was transferred photographically. Thus, photoset type and the use of web-offset presses were combined in an onslaught on the traditional and slow hot metal technology.

There are three components involved in the generating of cold type: the computer, whose memory for data storage is the starting point; the video display units (VDUs), by which copy is keyboarded into the computer and controlled and edited; and the phototypesetters which, by electronic commands, produce and deliver the bromides of the headlines and text.

The terminals

The VDUs, known more generally as terminals (Figure 1), are work stations consisting of a keyboard and a monitor screen, usually separated from each other so that they can be adjusted to the user. There are two main sorts of keyboards, those for writing and those for editing. Both have the traditional 'QWERTY' layouts found on typewriters, together with banks of command keys that enable the user to make use of the various functions in the computer.

On the writing keyboard, which has fewer command keys, the reporter types the story as on a typewriter, at the same time monitoring it as it appears line by line on the screen in front. A command key allows the story to be scrolled backwards and forwards so that it can be read through. There is also a facility for dividing the screen so that notes can be used on one half and the story entered on the other.

Words and lines can be altered and deleted by pressing command keys, the spelling and typing accuracy checked, and the length assessed before the story is 'sent' to the news editor's queue for checking on the screen, ready for editing. If the story requires further work, or has to be left while something else is done, it can be safely filed in the reporter's own electronic 'basket' until it is ready to be sent.

Each reporter has his or her own 'file' of stories, coded by name and catch-line, on which work is being done or is complete. The sum total of files comprises a 'queue' of stories – the newsroom queue, or the sports or features queue. These queues of stories, together with the agency queues, form the material of which the day's pages will be made.

The use of VDUs by reporters and writers is what is meant by the term *direct input*. Copy – the generic name for all material submitted for publication – is inputted into the computer by writers at the start instead of being inputted after editing. The advantage of this is that stories have to be keyboarded once only; once in the computer, copy can be processed through to the typesetting stage, so that the electronic impulse set in motion by the reporter becomes the keystroke that sets the paper in type.

Working outside the office need not prevent the use of direct input. Copy can be telephoned to copytakers who are using terminals, while small portable VDUs can be used at a distance from the office by telephone link-up to enter copy. There are various types of work stations for use outside the office. A device called a modem is used to attach a VDU to a public telephone line and copy is then simply dialled.

Editing terminals differ from the others in the extra command keys they have for editing procedures and the amount of access they have to the various sources of written material. By the use of commands, stories can be altered by deletion and insertion, the areas being worked on being defined by the cursor, or light pencil, which moves in response to the subeditor's key control. Use of the split screen enables two stories to be worked on together or to be merged into one, or for parts of stories to be rewritten alongside the

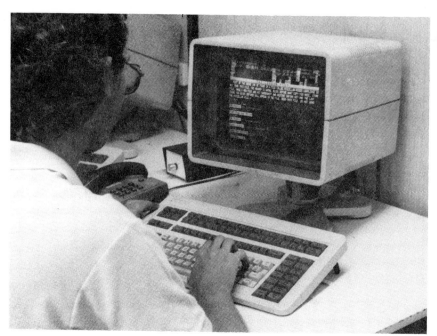

Figure 1 *Electronic editing being carried out on a VDU screen. Commands and alterations to copy are keyed in by the subeditor. The cursor, or light pen, responds to these and adjusts the text on the screen*

Figure 2 *Lines of type, or slugs, being set on a linotype machine under the hot metal system once universally used in newspaper production*

original. Grammar and accuracy can be checked and amended without difficulty on the screen and stories cut to length.

Headline types and sizes in common use, derived from type masters, are usually formatted inside the computer so that they can be identified by a single command. Another command will give a character count to show whether the headline fits or whether it is under or over measure, and by how much.

Finally copy is hyphenated and justified (H & Jd) by command key so that it comes up on the screen in the number of lines it will make on the page in the chosen type and measure, and with a word count. Thus the subeditor can see whether the story will fit the space or if it needs lengthening or cutting to fit. The story is then 'sent' to the photosetter.

The computer

At the heart of the system is the computer (usually with a back-up one in case of failure) whose memory and storage facility enables material for the newspaper to be gathered, stored, processed and turned into type. It is the development of fast computers that has enabled companies such as Atex, Linotype, Ferranti and Hastech to develop the all-in systems that not only enable cold type to replace hot metal but have revolutionized internal newspaper organization.

The modern computer allows work flow in newspaper departments to be organized, classified advertisements to be set, sorted, placed and invoiced, and rates to be worked out. The page dummy showing advert positions can be electronically generated, stocks of materials controlled and accounts automated.

In addition to editing and type generating for editorial use, some systems can offer access to a data base – for example, a 100,000-word dictionary check for spelling – and can sort and route incoming agency, or wire, copy to the right desk, and highlight local references for regional circulations. Complicated staff rostering can be worked out, internal mail and memos conducted electronically, and ongoing features such as stock market prices updated by the suppliers.

Photosetting

With the commands and the range of type masters available there is no reason, and no evidence, that the visual character of a newspaper suffers, or need change, through the use of phototypesetting. Provided it has been fitted out at the start with the types required, and has been properly programmed to deliver the sizes and measures needed, a photosetter is the useful workhorse of editorial production. It delivers what it is asked to deliver and at great speed.

While a photosetter can do the work of scores of linotype machines (Figure 2), the number used is crucial to the efficiency of the typesetting operation, and should allow for the increased workload of setting that occurs close to edition times. It is then that late stories, revisions, space cuts and changes in stories as pages are being made up can cause a tailback even in high-speed setters.

The problem is eased under direct input since this entails screen editing, which means that subeditors have continuous access to the computer and can edit copy electronically to fit. Where printing is by contract, or where the terminals and therefore the computer are controlled by separate operators, there is greater possibility of edited copy being over-set or under-set or needing adjustment.

The fewer alterations needed at the make-up stage the less risk there is of overloading the photosetters and endangering press times. As with hot metal setting, the burden placed upon the system at peak hours has to bear some relation to what is possible.

WEB-OFFSET

Photosetting was made possible by the computer but it owed its utilization to the web-offset printing process. The continuance of hot metal typesetting and page make-up had been unchallenged because it suited the plate-making process used in letterpress printing. The plates could be cast conveniently in metal from the moulds taken from the made-up pages. The plate makers shared the same hot metal and foundry facilities as the line-casters, which were located usually on the same floor. Moreover, metal plates in heavy relief had always been regarded as the only way type could be transferred successfully to newsprint by means of the high speed rotary presses used to print modern newspapers.

The main technical developments had been in the development and refining of the rotary press to make it faster and more automated. In typesetting an early use of the computer had been to improve the speed of hot metal line-casters by converting them from manual operation to a method using perforated tape. Those computer driven machines were able to lift setting speeds from five to fourteen lines a minute.

Improvements in the design of web-offset presses – web meaning reel-fed – in the early 1960s, which increased their print 'run' capacity and reduced paper wastage at start-up, was the breakthrough that was to make 'cold' or photoset type more than just a novelty.

Web-offset printing (Figure 3) uses a smooth zinc or aluminium printing

Figure 3 *A modern web-offset press in use at the News Centre, Portsmouth. The pages are printed not by direct impression from the printing plates, as in letterpress, but by being offset from a smooth polymer plate via a 'blanket' which picks up the page image in ink by chemical means and transfers it to the paper*

plate coated with plastic which takes on the page image by having its photosensitive surface applied directly to the page film under bright light. The surface of the plate is then developed with a chemical solution which creates areas that attract and areas that repel ink by means of a fine balance between ink and water. On printing, the plate image picks up the ink from a bath which it offsets on to a rubber blanketed cylinder which makes the actual contact with the paper, thus offsetting the ink impression on to the paper. The printing is thus quite different from the letterpress method in which the plate prints directly by impression on to the paper.

At first it was the chains of smaller papers that made the leap into cold type and web-offset, especially those that needed to modernize ageing plant. The switch was frequently recouped by the selling-off of highly rated centre sites and a move to smaller suburban locations, and a cleaner, quieter pollution-free work environment.

Bigger papers with heavier print runs and a costly investment in rotary presses, while looking to the cost-saving and other advantages of photo-setting, were reluctant to change to the new presses. Their reservation was that web-offset presses, while adequate for the smaller American newspapers, were not capable of delivering the bigger print 'runs' of British newspapers.

An early consequence, therefore, of the move to cold type was the adaptation of web-offset plate-making to produce impression plates for use on rotary presses. This was done by converting the page image to a polymer lower relief plate, which was then used to produce a mould from which a metal plate could be cast. To fit the presses, the plates were mounted on to metal saddles to give them the right thickness.

Thus newspapers that wished could dispense with hot metal in type-setting, block making and page make-up while still using their rotary presses, which might have a decade of life left in them.

An advantage was that the new plates were considerably lighter than the 18 kg conventional stereotype metal plate. A disadvantage was that, while the durability and cheap running of the rotary presses had been married to the cost-saving of cold type technology, plate-making time had been considerably increased and edition schedules upset by the extra procedures needed to convert pasted-up pages into metal printing plates. This was alleviated when the development of tougher polymer in the early 1980s, enabled traditional rotary presses to be used for direct printing from polymer plates derived from pasted-up pages.

Improvements continued to be made, however, in the design of web-offset presses in the face of increasing replacement of hot metal by cold type, and by 1986, the photoset the *Daily Telegraph*, at its northern plant in Manchester, became the first British national paper to use the new generation of web-offset presses which (they were able to tell readers) could print 60,000 copies an hour, the same as the big rotary presses.

This pointer to the future gained more potency from the fact that the new web-offset presses, while still slightly heavier on newsprint, were a generally cheaper capital investment. Also, they were more suitable for satellite printing centres where pages were being received photographically by page facsimile transmission from the main production centre and they were easily adapted for printing in colour.

DE-CENTRALIZED PRINTING

Page facsimile transmission is a technique whereby made-up pages can be photographed and transmitted from the main production centre to satellite printing plants within the circulation area so that the printing operation can be carried out simultaneously at several centres.

The page photograph is digitized by means of scanners so that the data is compatible for sending by microwave, for short distances, or by broadband telephone lines or earth satellite for longer distances. The signal is uncoded by a special receiver at the other plant and is made into a page transparency. This is fed into a plate-maker which produces a polymer printing plate identical to the one being used in the main centre. The average transmission time is about three and a half minutes.

The aim is to ease the distribution problem by locating the satellite centres at strategic points in the circulation area, thus minimizing newspaper delivery times. At the same time operating costs are saved by not having to duplicate the editing and typesetting operation.

The disadvantages are that extra editorial and typesetting capacity have to be provided at the main centre to process sport and other editorial material being used in the editions served by the satellite print. Also, pages carrying such material have to be specially created for transmission there.

An alternative to this solution, where readership in an edition area has to be provided for, is to have a hybrid operation in which the main pages are transmitted by facsimile and the edition area pages are set and composed locally. This is less cost effective since a nucleus of editorial, typesetting and make-up facilities has to be provided at the satellite centres.

Page facsimile transmission in America and Europe is invariably used as part of a computerized printing operation, but it can be adapted to hot metal production provided a special proofing machine is used which takes page proofs of a quality fine enough to be photographed. The transmission method can thus be used wherever there is a demand for it, irrespective of the printing technology employed.

This makes it an attractive proposition in countries, especially in the Third World, in which the use of satellite printing centres can be the answer to communication and distribution problems.

In Britain, to help in efficient distribution, a number of provincial and national newspapers, including *The Guardian*, *Daily Mirror*, *Daily Star*, *The Sun* and *News of the World*, are using page facsimile transmission for all or part of their production outside London, while *The Financial Times* and the *Wall Street Journal* circulate internationally on the same day by this means. Even *Pravda* uses the technique to reach the far confines of the USSR.

EFFECT ON EDITORIAL PRODUCTION

If we remember that printing technology is the servant of editorial production, it will help us to regard the use of computer-generated type in the same light. The aim of the new systems is to replace hot metal with something that will not involve any fundamental change in what newspapers are, or how they are edited and presented, but which helps them in their objectives. This applies whether the objectives are looked upon as a service to the community or as a means of giving shareholders an adequate return on their investment. The two objectives are not mutually exclusive since if one fails it takes the other with it.

Many of the changes that are referred to as new technology do not come within the area of editorial practice. Those that do do not alter the basic function journalists have in the production cycle. This is important to remember.

Copy provided by reporters and feature writers is selected and edited to the same standards and by the same principles. Deadlines are the same. Pages still have to be designed and made up, albeit with type bromides instead of metal or, in a few cases, by the manipulation of type and pictures on a screen after they have been edited in the case of photopage composition.

Computerized systems are designed to produce a newspaper that looks the same, or even better. If this is found not to be the case, then the wrong system has been installed for the particular paper (they do vary), or it has been installed wrongly, or it is being used in the wrong way.

There is one innovation, however, in that electronic aids are now available to production journalists. Where the new methods are being fully utilized, the reporters, for instance, still go about their work as before but instead of their copy being received in an indifferent typescript it is available properly corrected on a VDU screen. For the reporter has exchanged a typewriter for what is, in effect, a sophisticated word processor. Instead of overtyping words or crossing out lines with Xs, or even retyping the story, the reporter can amend his or her copy at a keystroke. Alterations of fact and correcting typing errors are performed with equal ease. The reporter with the minimum of keyboard ability will find, in fact, that a VDU flatters bad typing.

The subeditor, likewise, once the keyboard is mastered, can perform complex editing functions on screen involving the deleting and substituting of words, lines, sentences and even whole sections of stories, and the re-writing of copy on split screen against the original alongside. By the same means, revisions can be conveniently carried out by recalling a set story and amending it on screen and then re-running it through the photosetter.

Because of the keyboard's facilities the finished copy is clean, accurate, correctly spaced, and hyphenated and justified ready for entry into the photosetter, where it will be in no danger of mechanical error. There is no need any more for subbed copy to be a mass of crossed out words, connecting lines and bits of small writing between the text.

This electronic new world depends, of course, on the use of direct input of copy and screen editing. In offices where this is not the case, certain variations might apply. Control of the terminals by printer/operatives under work agreements can mean copy being inputted only after the subbing stage, when it is ready for setting. In such cases subbing is carried out on typewritten and teleprinted copy as under pre-computer systems. In other offices copy might be keyed in at the outset by printer/operatives rather than reporters, with subbing and setting commands being done later on terminals by journalists. In the case of contract printing the typesetting operation is often geographically removed from the writing and editing side, with edited material being carried from one to the other, and proofs or print-outs for cutting and amending being returned from the print centre.

With the movement towards direct input and editorial control of terminals as the human problem of shrinkage and redeployment of the old printing workforce is resolved, many of these variations must be regarded as interim ones. The ultimate logic, as the American experience has shown, must be direct input if the fullest utilization is to be made of the systems that have replaced hot metal. All American newspapers now have direct input.

In our examination of editorial production that follows, account will be taken of the challenge of the new systems just described and the utilization of electronic aids by journalists in putting a newspaper together.

2 CHOOSING THE MATERIAL

The production sequence on a newspaper begins with the selection of the contents for the day's issue. This does not mean that the writing part has finished by then. News will continue to be gathered and written up to the last deadline of the last edition. But once decisions begin to be taken about the placing of stories in the pages, then that is production.

Since 'copy' is the term for all material, whether handwritten, typed or keyboarded into the computer, copy-tasting is the word used for the selecting of this material, and the person who initiates this process is the copy-taster.

News creation is a round-the-clock operation into which the newspaper tunes during its hours of production. On a British evening paper, where work begins at about 8 a.m., the newsroom deals first of all with overnight stories and follow-ups from the previous day's news. Because of the time scale, foreign copy will at first be predominantly from America, and so the early editions of evening papers often carry American-originated material which is later thrown out as the flow of home and European stories picks up with the start of the day's activity cycle.

As edition succeeds edition from about midday there is a strengthening of the home-news-of-the-day content, with running stories being updated between editions into the late afternoon, the peak time for the in-flow of stories being around 2 and 3 p.m. Afternoon cricket scores replace morning and overnight ones. The times of racing results, the later ones carried in the stop press or late news box as the afternoon draws on, are a guide to the press times of the edition that you buy.

The use of a late news box (sometimes called the fudge) is important to an evening paper, for Parliament, the courts, sport and other news events are still running in late afternoon. Brief summaries which would otherwise have missed the edition can be inked in with a special printer during the actual printing run. Late evening press times seldom go beyond 4.30 p.m.

The morning paper work cycle begins at about 11 a.m. or midday with edition press times from about 7 p.m., depending on the distribution areas. National papers printing only in London print earlier for distant areas, although the growing use of page facsimile transmission is making possible the simultaneous printing of a paper at a number of centres, so that readers in distant areas do not need to suffer a lack of late news. By comparison, provincial morning papers, with their tighter distribution area, can afford to work until well into the evening before getting their first edition away.

Morning paper newsrooms first of all mop up the news of the day, looking for follow-ups where stories have had a good run in the evenings and on television, and giving more detailed versions of stories that the evenings have only been able to nibble at. They have the benefit of evening events, speeches, meetings, functions and theatres, and have more time for set piece interviewing.

Because of the time sequence, morning papers are stronger on European news and are first with the day's events from Australia and the Far East. They also have time for polished detailed background features and situationers which give depth and perspective to their coverage.

By the time the last bit of updating has been done and the last edition (not usually later than 2.30 a.m.) has gone to press, the flow of home and European news has subsided and the first stories are coming in from America. The news cycle is ready once more for an 8 a.m. start-up by the evenings.

NEWS SOURCES

This input of news arrives in a newspaper office from a variety of sources and by a variety of means. The means, in particular, have undergone changes in recent years with the use of the computer. It is now possible, as we have seen, for reporters to keyboard their copy straight into the office computer, where working arrangements have been agreed, and for news agencies to do likewise. Stories entered this way are recalled on to screens for editing before being typeset. Agency copy entered into the computer can be automatically routed to the desk whose job it is to read and deal with it. Local references in agency stories can be type-coded to local subscribers at the keyboarding stage, so that the copy-taster's attention is drawn to them.

This is the sophisticated end of news input. With many of the news jobs being carried out by reporters in Britain and around the world, however, the telephone, the ballpoint pen and the typewriter are still means by which news copy is initiated.

Staff reporters

Staff reporters are the most useful and controllable source of news stories, and the likeliest source of exclusive material. Their numbers vary from about twenty to thirty on a small evening paper to forty to sixty on a national daily. They operate from the newsroom or from branch offices, and sometimes independently in districts, and are controlled and briefed by the news editor. Briefings can vary from a simple inquiry by telephone or personal call arising out of a letter or information, to complicated jobs involving a team of reporters and sometimes several locations. Some stories can have both home and foreign 'ends' and last all day or several days.

Reporters' copy is telephoned from locations when edition deadlines are pressing. If it is a long running story it is telephoned in 'takes' to meet each edition. If there is time it is typed or keyboarded when the reporter returns to the office. Telephoned copy is taken down by telephone typists or keyboard operators, who are called copytakers (in the US, telephone reporters).

Where there are bigger staffs a newspaper might use a number of reporters as an investigative team on long-term news projects under a leader or editor who collates the material for use. This is usually planned into the paper in advance as a special news feature. This type of in-depth journalism was pioneered as regular content by the Sunday Times Insight team in the 1960s although a number of Sunday papers had been doing it for some years on an *ad hoc* basis.

Freelance reporters

Freelance reporters are used mainly on specialist assignments or for holiday relief work and are paid per job or by per days or weeks worked. They are not tied to any paper except where they have a specific contract covering a job or sequence of shifts. Some specialize in types of investigative writing which they sell exclusively to the highest bidder, while others rely on a variety of arrangements with newspapers or magazines, often specializing in a subject for which they become known to editors and readers.

Local correspondents

Local correspondents are journalists mostly working for a local paper, though they are sometimes freelances, and they are accredited by retainer or special arrangement to offer their services, when needed, to a bigger paper circulating in the area. They are paid lineage (a fee per line) for stories used and for stories ordered even if not used. The arrangement usually precludes them working for a rival bigger paper, though they can work as correspondents for the BBC or ITN (Independent Television News) or for publications other than newspapers.

Lineage payments are a useful source of income for country and small town journalists. These are a valued part of a newspaper's news network. They cover stories for which a staff reporter is not normally available.

News agencies

News agencies are vital to the newspaper industry. National and international agencies work round the clock to provide a variety of services for newspapers all over the world, collecting material from bureaux and correspondents in cities and countries, checking and editing and then distributing it.

Most countries now have their own national agencies whose services are especially used by small- and medium-sized papers to fill gaps in the coverage, or even to provide all but local coverage. Agency stories can be used as check sources, and many agencies also provide news pictures. International agencies such as Reuters, Associated Press, United Press International and Agence France Presse are the prime source of foreign news for newspapers with few foreign correspondents of their own, and also of newsfilm and sound reporting for television and radio. They feed into national news networks stories affecting a country's own interests and nationals abroad. Reuters, for instance, do this through a series of regional services to the Far East, Western Europe, North America, West Africa, etc. News agencies are often the first to break important foreign news stories.

They operate in a similar way to newspapers through staff reporters in main centres with local correspondents, or stringers, filing in from the districts. The subediting is a round-the-clock shift operation serving morning, evening and Sunday papers.

Most newspapers in Britain take the Press Association (national) and Reuters general services and some the sport, financial, situationer and other services, depending on their contract. A fixed rent is paid for each service, whatever is used. Subscribers take agency copy either through the traditional teleprinter machines, which unroll news copy as they receive it, or straight into the office computer where it is stored to await tasting and subediting.

Hand-outs

Hand-outs arrive in newspaper offices in great numbers from all sorts of organizations, including the Government, and sometimes from celebrities, via press officers whose aim is to reach the public through the newspapers. They are usually given to reporters to read or to specialists whose field they cover in case they contain something that will be of interest to the readers.

Reading hand-outs can be a tiring, and sometimes boring, job but they can yield important stories over a wide field including, for example, housing statistics, immigration, technology breakthroughs, new cars and copy for consumer columns.

THE NEWSROOM

Sociologists have divided newspaper journalists into two sorts: the gatherers and the processors. The gatherers, if we are to accept these useful terms, are the ones who find the news and put it into words. The processors are the ones who turn the words, and also the pictures, into a newspaper. This is the part of the operation with which this book is concerned. Yet, though the newsroom is the heart of the gathering side of the business, it is concerned with the production side, too, for it is here that decisions are taken that will influence the news content of the paper.

The news editor (sometimes the chief reporter) who is in charge of the newsroom, is one of the busiest and most important executives on the paper, for the quality and extent of its news coverage depends on the daily operation that he or she puts together. The assessment of a news situation depends at the outset on the news editor's judgement: what news is to be covered, in what depth, and who is to do the work. Reporters and correspondents are briefed so that copy is filed on time and important aspects of stories not missed, stories are commissioned from freelances and outside specialists, the agency input is monitored, work is checked to see that it has been properly done, and hand-outs, correspondence, expenses, travel arrangements, interviewing of staff and a hundred and one things are dealt with. It is all aimed at producing a satisfactory flow of copy for the news pages. In this work the news editor is helped by a deputy and sometimes by one or two assistants.

The news editor has to keep a check on where the staff are and decide when to extend coverage of a story, when to ask for more copy and when to call it a day when a story has 'fallen down'.

The use of computerized systems has meant the dawn of the *electronic newsroom*, with news editors checking reporters' files on the screen, sending back stories where coverage is not sufficient or has failed in some way, and routing stories to the copy-taster and the subeditors as they become ready.

While stories are still telephoned to copy-takers, the use of portable VDUs enables reporters to key their own stories into the office computer direct from location, linking their VDU through the public telephone system. District offices can likewise be controlled electronically through the newsroom's keyboards, with stories and messages being routed to and fro.

Village and suburban correspondents of provincial papers are beginning to have their own portable terminals, where the volume of copy warrants it, so that Women's Institute reports and flower show results are creeping into the computer age. The *Portsmouth Evening News*, which now has a number of such terminals in use among its part-time contributors, has also had private word processors linked to the office computer where they can be used for copy origination.

The result of these new methods is a speeding up in the transmission and handling of copy (the days of the old 'train envelopes' are gone) and a more effective control of coverage by news editors.

Thus, whether by electronic or traditional means, by the time the copy-taster begins work, a good deal of decision-taking has already taken place. The material has passed through the newsroom system. Among it, the likelihood depending on a fine equation involving the facts of a story, the volume of space available and the volume of news about, are the candidates for the day's pages.

THE COPY-TASTER

In theory the editor reads everything that is likely to get into the paper. In practice this is not always possible, although he or she is usually given a copy of all reporters' stories, or can call up material electronically on to a personal screen.

The editor, through informal discussions that take place with executives during the day, and through the briefings and debate of the daily planning conferences is familiar with the more important stories that are being covered; at an early stage in the production cycle, the editor has ideas about the paper's thrust and news balance. The likelihood of some big stories is known in advance and their place in the paper can be prepared. Nevertheless the editor would probably not, for instance, read all agency copy unless looking for something in particular.

There are also days, when, because of other duties, he or she will have little time to focus much on the news input. The role of the copy-taster who, come what may, reads everything that is written and submitted, is thus an important safeguard.

The copy-taster is usually the first journalist to arrive for duty in the editorial room. Here, at a desk close to the chief subeditor (Figure 4), the process of 'tasting' goes on until the last edition of the day has gone to press. The work is generally split into two shifts with a deputy taking over later in the day, or the two alternating shifts from day to day.

20 *Modern Newspaper Editing and Production*

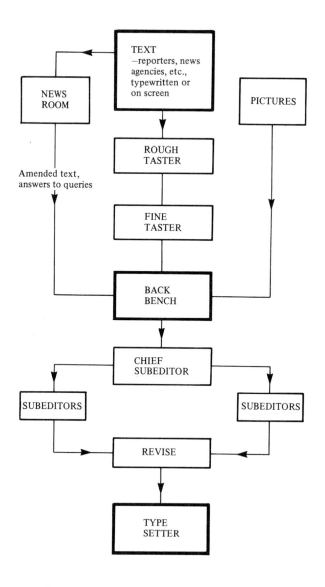

Figure 4 *Copy-tasting: the sequence of copy flow on a big national newspaper*

This journalist figure is the one most maligned by the media sociologists. The copy-taster is the 'gate-keeper', the one who lets in or shuts out stories according to the line of indoctrination being practised by his proprietor/organization; who assesses news to a set of stereotypes deeply embedded in the professional soul; who bends stories to fit pre-conceived categories and pigeon holes; the stopper when anything original or damning to the system appears at the gate; the one who protects the readers, the people, the organization, from truths that jar . . . and so on.

The job, in reality, is nowhere near as influential or as sinister. The copy-taster is unfailingly a senior journalist who understands the paper's market and readership, and the requirement is to sift through and sort incoming stories and to reject those which, for various reasons, are unsuitable or unusable. The result is to short-list and reduce to manageable proportions what is sometimes a mountain of material, so that the night editor, chief subeditor, or whoever is in charge of page planning, can set to work without being swamped by stories and their time wasted in reading.

The danger facing all copy-tasters is of 'spiking' a story which turns out to have been one the editor feels should have been carried. This is a special hazard on national papers where the rejected story might turn up in a rival paper. The safeguard against this happening is the copy-taster's own finely tuned news sense and familiarity with the paper's style and content.

MEASURING THE NEWS

A quick examination of a number of newspapers on the same day will reveal that while some stories are common to many of them, there is a great variety of opinion among them as to what news to carry. First of all there is a difference in news coverage in the daily press between newspapers that circulate nationally and those that are regional or local. Within these broad bands there are further differences in market and content.

Some newspapers exist to propagate the views of political parties or of churches or minority groups or interests. Some national papers – in Britain, for instance, *The Times*, *The Guardian* and the *Daily Telegraph* – have a serious, or 'up-market' readership interested in cultural things and world news, and in technology and education. Some are highly specialized, such as the *Financial Times* with its business news. At the same time many of the tabloid, or half size, papers (which have a bolder display) such as *The Sun* and the *Daily Mirror*, sell to a popular readership and are strong on pictures, human interest news and stories about TV, show business, sport and the Royal Family.

Local town evening papers reflect the life and activity in their circulation areas and carry only the more important national stories and mostly very little foreign news. Some country papers are full of news about the farming community, village activities, local weddings and county social gossip.

News about dominant industries, such as steel or textiles, in which many readers work, can feature strongly in regional papers. This is all quite natural. They are performing a service for their readers.

There is also a difference between daily and Sunday papers, and between provincial evening and weekly papers. The once-a-week publications devote more space to features giving depth and background to the news and to gossip and opinion. Some Sunday papers have an intense preoccupation with the politics and economy of Britain and other countries.

It will be seen from this quick survey that there can be no universals in news assessment. While the job of a copy-taster on a national 'serious' daily and that on a small provincial evening follows the same routine, the yardstick by which they measure what stories are worth using is quite different most of the time. There is, in fact, no yardstick by which a copy-taster can identify news worth using which holds good for any newspaper.

This does not mean that there can be no definition of news. News remains, in any circumstances, the first tidings, or knowledge or disclosure of an event.

A secret or unknown event remains secret and unknown, for instance, until upon its first disclosure it becomes news. In the old days of horses and carriages, word about a great battle might take up to three days to cross Europe to the office of *The Times* in London. But it is only then, when it is disclosed, that it becomes news when, as a nineteenth-century editor of *The Times*, John Delane, once said, 'it becomes the property of the nation'. In the same way, the marriage in secret of a celebrity might take ten years before it is revealed to the public. It is then news.

Thus we can say that, in terms of a newspaper's content, it is not the event itself that is news but its disclosure.

Equally, once a situation has been disclosed and made public it ceases to be news. If a notable news story that has attracted great attention appears in the paper one day, it will be referred to the following day on the news pages only in the light of some new aspect that has cropped up. This is what is meant by a *follow-up*.

Yet it is not sufficient for an event to be news within this narrow definition for it to be worth printing. It must also be of interest to the potential readership which, as we have seen, can vary from paper to paper. There is little point, for instance, in a country weekly on the Welsh Border filling its pages with news about Brazil or China, or the *Financial Times* concerning itself with the result of the three-legged race in a school sports in a Yorkshire village.

The journalists involved in editing and producing these papers know their market – they have to know it to do their job properly – and they select the sort of news that is right for the paper. One of the jobs of the newsroom is to see that unsuitable stories are not worked on. Even so the copy-taster will find that among the news that flows into a newspaper office each day there will be a percentage that is unlikely to be used because it does not fit the pattern. The first job is to weed this material out.

A more potent factor in copy-tasting can be the sheer volume of news available. A newsroom or a news agency does not stop filing when it has reached a given number of stories; it goes on covering the news wherever it breaks. The quantity of stories can vary from day to day but, as far as the gatherers are concerned, it is the job of the editor and his or her production staff to decide how much of it will be used.

While the factors that govern this measuring of news become almost instinctive in an experienced copy-taster who knows the paper, it is worthwhile noting down what they are. Also, it should be repeated that it is not the copy-taster's job to select what goes into the paper but rather to exclude what is unsuitable, short-listing the remainder into a manageable amount.

A story is rejected if:

1. It is geographically outside the range of the paper's market, unless the story is of special importance (depending on whether it is a national, regional or local paper).
2. It is outside the readers' range of interests (depending on whether it is a quality, or up-market, paper, popular or specialist).
3. It does not extend any further material that has already appeared in the mornings or evenings.
4. It appears to be merely seeking publicity for someone and has otherwise little reader interest.
5. It is legally unsafe, in bad taste, racist, clearly inaccurate, silly or based on rumour.
6. If it is simply not good enough or interesting enough on a day when there is a great deal of news about.

The last point demonstrates in a telling way why absolutes are not possible in measuring news value. Even if a story fulfils most of the criteria, the decision on whether it is used and, if used, how much space it will get, can depend on the number of good stories demanding to be used on the day, the existence of other stories of a similar sort, or the lack of space due to a shortfall in advertising.

It is not generally realized that the amount of editorial space available in a newspaper, as a rule, rises and falls in response to the flow of advertising and not to the volume or quality of news available. The consequence of this is that a story that might justify being given a good space one day might

Figure 5 (a)   *A typical busy town evening news page – the* Evening Argus *at Brighton, Sussex*

warrant less on another. This can happen on certain days when the copy flow is slack – on the Sunday shift working for a Monday morning paper for example.

The alternative on such days might seem to be to leave blank spaces on the pages, but this would look silly and also would not be liked by the reader. What happens is that certain exclusive stories are held back if possible (or planned forward) for publication on slack days, a greater number of foreign

Figure 5 (b) *The* Brighton Argus *page seen as a pattern of highlights created by headlines, pictures and advertisements*

stories used and more space given to features or perhaps picture display. It has to be said that, however uneven the flow of news, there are very few days in the year when the editor is at his or her wit's end what to put into the paper, but the phenomenon just described does account for the good and bad days that newspapers have, which are sometimes noticed by readers.

PATTERNS OF NEWS

The pattern of news selection can be seen at work in the examples of news pages reproduced in this chapter. Brighton's evening paper, the *Evening Argus* (Figure 5), sticks closely to the edition's circulation territory – in this case Sussex – to the north and east of the town. The lead story concerns Hassocks and Haywards Heath; the half lead across the top of the page mentions Hailsham and Uckfield. The picture subject and the filler down column one are also from Hailsham. There are stories from Lewes and Brighton Crown Court, and a story in column one that concerns East and West Sussex. Unusually, only one of the three advertisements is local. While the local reference is given in the first or second paragraph in all stories, none of the headlines has local geography in it. It is an edition page on which the

Figure 6 *An active news page of the* Western Mail, *a morning paper covering a wide provincial area*

readers expect to find the material relevant to their area. It does not have to be pointed.

The broadsheet Cardiff-based *Western Mail* (Figure 6) with the wider regional catchment area that provincial morning papers usually have, justifies its claim to be 'the quality voice of Wales' by carrying a broad perspective of Welsh interest news. The lead looks at the effect of unemployment and new jobs in the country as a whole, while the remaining nine stories come from various parts of North, South and Mid-Wales and range from a police chief's retirement to a battle to save a village from being ruined by a rubbish tip.

The large page, with its one small advertisement on the bottom left, is structured around a picture spread in which a Welsh conference of the Twins and Multiple Births Association is used as an occasion for presenting two attractive pictures of child twins.

The choice of story, with its avoidance of geography in the headlines and its variety of subject, shows that the page is aimed for acceptance by readers in all parts of the paper's circulation area.

The news page from the national tabloid daily *The Sun* (Figure 7) shows entirely different considerations in its news selection. The lead story, SOCCER ACE'S GRIEF AS WIFE DIES AT 28, is a bold presentation of a poignant human interest story that derives nothing from its geography. The half-lead, MURDER RIDDLE OF KIND OLD LADY IN LOOTED FLAT, is in the same category. The geography is not given until paragraph four and is not significant to the story. Of the remaining three items on a page with an uncomfortably large advertisement, the biggest is a story of a strange killing from New York, the others an item about a wet-weather drought in the West Country and a small panel about a jet export order to Oman.

The effect is of an active news page in which items have been chosen for strong general reader interest, coupled with a bold presentation aimed to counter a dominant advertisement, a problem that neither of the other two examples of news pages has to contend with.

The final example, page four of the broadsheet national daily, the *Daily Telegraph*, exemplifies the wide news coverage in the 'quality' newspaper field. This is one of two pages given wholly to overseas news in a 24-page paper that also has overseas news mentioned on four other pages.

Out of eleven stories on a page which is nearly half advertising, the following countries are covered: Italy (the lead story), Pakistan (the half lead), Austria, Belgium, the US, Sri Lanka, Israel and the USSR. Only one is an agency (Reuter) story, which shows the importance the *Daily Telegraph* attaches to the contribution by staff reporters and correspondents abroad. Geography, either by place or name, appears in nine of the headlines, which shows that the stories are aimed at readers who are looking for world news coverage.

Figure 7 (a) *How a popular national tabloid paper,* The Sun, *presents news in a more visual way*

Figure 7 (b) The Sun *news page seen as a pattern of highlights created by headlines, pictures and advertisements*

Figure 8 *The news treatment of a traditionally presented national morning paper, the* Daily Telegraph

TASTING METHODS

The effect on news copy-tasting of electronic copy inputs is more apparent than real. Whether on screen or on hard copy the basic routine of the job is the same. Important stories are drawn to the attention of the night editor or chief subeditor; the clearly dead ones are spiked, the doubtful but possible ones put into a separate pile to be turned over in moments of need, the likely tops of pages put aside for use as page planning proceeds.

Those that might make intermediate tops of pages are put separately by. Stories that are linked in some way are kept together, while a flow of shorter but useful material is fed to the chief subeditor for use as fillers on pages. These are usually subedited throughout the production period so as to be available for spaces on various pages and editions and can be stored in the computer in photosetting systems.

The copy-taster keeps a look-out for stories with edition area interest and brings them forward at the right time. Agency foreign stories are referred to the newsroom for possible home 'ends' where this has not already been done. Stories with useful news points buried deep down in the copy, which might otherwise have been spiked, are marked and shown to the chief subeditor or sent back to the newsroom for inquiries. Notes are sent to the picture desk of any picture ideas suggested by copy. With agency copy, electronic inputting can short-cut the work, as we have seen, by routing the copy to the right desk and type-coding local references.

Some national papers refine the tasting process by filtering copy through a rough and a fine copy-taster, or through separate home and foreign tasters. The aim is not only a fail-safe reading operation but a continuously creative assessment of the copy flow to pin-point things that might otherwise be missed.

Electronic copy-tasting

Electronic copy-tasting, in which the taster reads from the screen, is necessary once direct input of copy is adopted – when reporters and correspondents type their stories directly on to a VDU instead of a typewriter, or when their telephoned copy is likewise entered into the computer by the telephone copy-takers. Once such copy has been cleared by the newsroom it can be recalled from its 'queue' in the computer for tasting and, after that, for page planning and editing.

Electronic copy-tasting is made easier not only by the taster being able to call up stories at a stroke, but by being able to call up the complete 'directory' of stories held in a given queue (i.e. newsroom, agency, sport, etc.) which gives the source, name and catch-line and the first few lines of each story, and also its length.

Stories can likewise be sorted into subject and priority files within the computer so that the right material is drawn to the chief subeditor's attention at the right time. Queries can be keyed back to the newsroom if the taster feels that a story needs further work on it, or to inquire if pictures are being provided with a story. Agency stories with local angles of use to the paper can be queued together; stories for use can go to the 'pending' queue; unwanted stories can even be 'spiked' by pressing the appropriate key. An all-electronic copy input thus makes the job of the copy-taster simpler. Even so, on heavy news days, 'electronic' copy-tasters find it useful to reinforce their screen data with a few written reminder notes.

Yet with direct input there are still, on every provincial paper, some part-time journalists and village correspondents who do not normally have access to a VDU, and submit handwritten or typewritten copy. In offices this is now usually keyboarded into the system by the newsroom after being checked and given basic editing.

One of the problems is that the introduction of computerized systems has led to a number of different production practices as companies have grappled with the task of reducing and retraining their traditional print work force. In offices which have opted for an interim stage in which keyboarding is done by traditional print workers after editing, copy-tasting follows its traditional hard copy pattern.

It is not uncommon for hybrid arrangements to exist in which reporters' copy is inputted directly, and other copy is accepted in its typewritten form, on which it is subbed before being keyboarded. Such systems are not satisfactory from the copy-tasting point of view, with tasting being a combination of paper and screen work.

The different stages which computerized technology has reached in the various production centres has resulted in news agencies – the principal suppliers of news copy in any country – having to maintain a variety of services to satisfy all their subscribers.

The Press Association, Britain's national news agency, which is co-operatively owned by the principal newspapers in the country outside London, and of the Irish Republic, transmits on seven channels. Two are home news channels, one is for sports reports, one contains a digest of Reuters, Associated Press and Exchange Telegraph material for provincial use, one has fast sports' results, another carries pictures, and the last one a computerized service, including computer formatted material such as race cards, with a facility for updating.

By using the new network of fibre-optic cables instead of telephone lines, the agency digitizes all its transmissions, including pictures, through a processor called a message switch. This feeds them into a decoder at each production centre. This is a device that sorts the signals and pushes them into a microprocessor which is programmed to select the appropriate service.

The subscriber makes the choice: text or pictures, computer input or teleprinter input, general service or private wire. Thus the use made of the agency's transmissions is decided upon, and modified as necessary, at its destination.

The aim of the system is to rationalize editorial production by enabling each centre to take what it wants to suit its own system. From the copy-taster's point of view there is the option of electronic tasting of agency copy where full direct input is in use, or of tasting the same story on traditional teleprinter copy where it is not. It provides the means of moving easily from one to the other as systems are updated and extended.

The essential thing with electronic copy-tasting, as it becomes the norm, is that the basic function as outlined earlier should not be allowed to be frustrated or diverted by the switch to a different copy input system. The copy-taster's role remains a vital part of the production cycle, and the electronic storage and retrieval of text does not diminish the role.

3 PAGE PLANNING AND TYPOGRAPHY

The plan of a newspaper – what it contains and the order in which it is presented – is decided by the editor in consultation with his or her senior executives. It is usually initiated at a formal editor's conference held early in the day on a morning or evening paper, or early mid-week on a Sunday paper and thereafter adjusted, if need be, in the light of later news and pictures.

The editorial space which, as we have seen, can vary from day to day in response to the volume of advertising, is shown at the start of each day or week in a blank dummy of the paper prepared for the editor by the advertising department. This contains the number of pages decided by the management, together with the spaces and sizes of advertisements sold.

The advertising content, over which the editor has no control, thus provides fixed points around which the editorial content of each page is planned. This does not mean that the placing of advertisements is entirely arbitrary. As a rule the front page is kept fairly clear, often completely clear, and other key pages, including the back, are kept 'light' to suit known editorial needs, but within these loose limits the number and shape of advertisements can vary a great deal.

A small amount of 'trading' of space might be possible between the editorial and advertising departments but, on the whole, advertising, once placed, cannot be altered. Many positions have been specifically bought by advertisers at a premium. Advertising also has to keep its own balance in the paper to avoid product clash or having cut-out reply coupons on pages that back on to each other.

FORMAT

Nearly all newspapers develop a visual format which is distinctively their own, and by which they are recognized, and the daily planning adapts to this. By format we mean the consistent use of the same typography and style of presentation and the placing of things such as sport, editorial opinion, women's pages, TV programmes and late news in the same part of the paper in each issue so that the paper has a shape which is familiar to the reader. This familiarity is important in page planning.

The format of a newspaper is its brand image and while it can be modified in various ways, both in content and in typography, it is not usually subject to drastic changes unless the paper is being relaunched in the face of falling readership, or is deliberately seeking to change its market.

A characteristic of the format of some national papers is that foreign or financial news has its own part of the paper. Some, especially the quality Sundays such as *The Observer* and *The Sunday Times*, sectionalize their contents in the American style, sometimes with their own pagination and masthead logo. Sectional format is not used very frequently with the smaller provincial and local papers, although many reserve a set part of the paper for pages of news aimed at edition areas.

Within these formats, however, it is still necessary to achieve a balance of subject on the pages and also a visual balance of text, headlines and pictures in relation to each other. It would be unbalanced, for instance, to run all stories of gloom and doom, or of death and destruction, on one page, to put all human interest stories on the same page, or to fill a page with stories without pictures. Account is taken of the advertising content on a page so that there is not a visual or contents clash with editorial material.

The editor might also occasionally vary the balance of space allocated to news, features and sport to suit special circumstances – an election edition or a heavy sports programme.

Each newspaper arrives at a general contents balance to suit its market. Evening papers give more space to news than Sunday or weekly papers. This is because they are exposed to the main daily news breaks whereas Sunday papers, produced on Saturday nights, mop up and explain the background to the week's news and run more weekend leisure features. Morning papers, too, tend to have more features than evening papers because of their tradition of giving in-depth explanation to the day's news. However, newspaper markets are by no means fixed and there can be subtle shifts in contents balance and news coverage over a period.

No matter how the balance works out it is achieved within a total editorial space which remains at about the same percentage of the whole in relation to the advertisements in each paper, whatever the number of pages. Advertising is at its lightest, at around 30 per cent or less, in town evening papers, which have lower overheads; at around 40 to 45 per cent in national popular dailies, which have high overheads but a high circulation revenue; and 60 per cent or more in the quality Sundays, which have high production and distribution costs coupled with less circulation revenue.

We can say, then, that the essential points in planning a newspaper are:

Page planning and typography 35

1 Balance of editorial content in relation to each other, i.e. news, sport and features.
2 Balance of subject content, including pictures and advertisements, within each page.
3 Preservation of the general 'shape' of the format.

WHO DOES WHAT

Editorial manning levels vary according to the size of a newspaper. Big national dailies have a battery of executives, some with precise editorial functions, some with managerial, administrative or liaison roles, some with general overlord responsibilities for certain areas. While there is usually only one deputy editor, who is the executive who actually deputizes in the editor's absence, there can be a number of assistant editors with titles such as assistant editor (news), assistant editor (features), assistant editor (special projects), etc. At departmental level (something akin to middle management) there are more defined roles. The titles of news editor, features editor, and sports editor mean that they are head of these respective departments, each with a deputy (Figure 9).

The production cycle in big dailies and Sunday papers is the responsibility of the night editor, who is the senior production executive, usually equal in status to the assistant editors. To the night editor are delegated the actual editing functions of the editor once the departmental heads have gathered the material and the production cycle begins. From the night editor flow the delegated roles of the subeditors and page layout artists, via the chief subeditor and the art editor. It is the subeditors and layout artists who do the detailed text and picture editing and page artwork.

In a small provincial newspaper, on the other hand, there might be only an editor, deputy editor and chief subeditor, plus news editor, features editor and sports editor (Figure 10). Similarly scaled down is the team of subeditors who might combine the news and features functions and even that of sport, and also attend to picture editing and the drawing of pages. There are no separate page layout artists.

Whatever the scale, the role of the chief subeditor stands out as the fulcrum of the production operation. Into his or her basket (wire or electronic) land the copy, pictures and ideas that have been ordered, discussed and provided, and out of it flows the finished material, checked, edited and prepared for typesetting or camera, that will make up the day's editions. With the page layouts in front, the chief subeditor works through the material of the day, page by page, edition by edition, briefing subeditors on the handling of stories, accepting or rejecting headlines, solving the problems that arise and sometimes even checking the finished work. In this

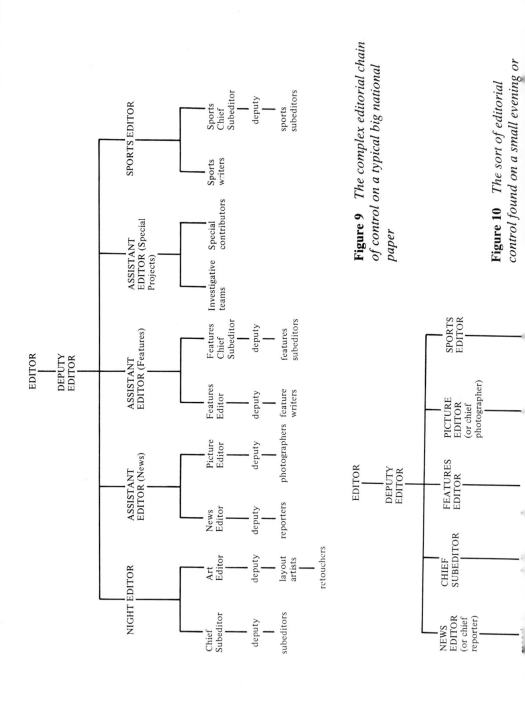

Figure 9 *The complex editorial chain of control on a typical big national paper*

Figure 10 *The sort of editorial control found on a small evening or*

the chief subeditor is helped by a deputy and sometimes an assistant, one of whom might take over the job of checking or revising the finished work before it goes for typesetting, or sit in as chief on certain days. The revise function is particularly important with direct input and electronic editing, where there is no traditional proof reading room. The revise subeditor must ensure that editing errors are spotted and house style adhered to.

The job of the chief subeditor combines a piece-by-piece planning operation with quality control of the material that is being processed through. It is an arduous and unrelenting task. On a small paper it might also include the scheming of many of the pages and a major share in the decision-taking about the balance of the paper. In a big national paper with the 'back bench' system, this aspect of the work is covered by the night editor and assistants, leaving the chief subeditor to give greater time to the polishing and quality control of the material.

PRINCIPLES OF DESIGN

Once the plan of the day's paper is agreed, the presentation of the contents becomes an exercise in design. This is called 'scheming' the page. The page plan is called a layout.

On a big morning paper the night editor and assistant night editor scheme the main news pages, with the chief subeditor perhaps scheming some. On small papers, as we have seen, the chief subeditor plays a bigger part, with some pages being given to senior journalists to scheme. On bigger papers, layout artists, under the direction of an art editor (on magazines, the art director), draw the pages in detail after they have been schemed, and prepare any artwork such as headline-picture composites (compos), logos, motifs and reverse type headlines (for example, white on a black background).

Whatever method is used, the page executive produces a layout which indicates the type setting, space and length allocated to stories, the exact sizes and shapes of pictures and the type and size of headlines. The layout then becomes the blueprint to which the stories are edited and from which the page is made up (Figure 11(a) and (b)).

It is possible to formulate some general principles about page design without contradicting what has been said about format or stultifying the use of new ideas. But first it should be said that, to succeed, the design must project successfully the sort of content in which the newspaper specializes to the sort of market that it seeks. In other words, content governs projection. Design, therefore, must look to the content and readership market for its inspiration.

To give some examples: the job of a local evening paper (Figure 5) is to

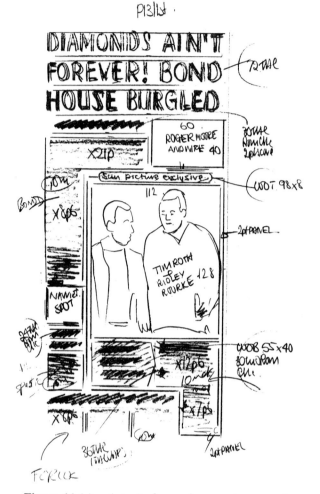

Figure 11 (a) *A typical page layout plan, or scheme – in this case of a news page in* The Sun *– indicating headline and picture sizes, and position and length of stories*

provide a comprehensive news coverage in its area so that residents will regard it as *their* paper and be attracted to buying it or to continue buying it. The design must give an impression of busyness, with pictures that connect with the reader, but also with a good space on each page for news, a choice of informative headlines and a range of topics which will serve as wide an area of local interests as possible.

A popular national tabloid paper (Figure 7), half size for ease of handling, is read at odd moments during the day by people going to work or in lunch

Figure 11 (b) *How the same page appeared when printed*

breaks, and by busy housewives, rather than dwelt upon in the evening. It is more likely to be bought on impulse than be delivered, and it has a younger readership than other national dailies. Many of the readers have little interest in a blanket coverage of events and are seeking rather to be entertained or stimulated. They are looking for human interest stories, eye-catching pictures, a favourite comic strip, and probably the sports gossip. The result: short news items, big pictures, exclusive stories about favourite subjects such as the royal family and TV characters, invitations to win money, and a brash style of headline writing. Such a newspaper is determined to be a survivor in the news-stands and its bold typography leaps out at you.

Take a specialist daily: *The Financial Times* (Figure 12). Here is a paper

Figure 12 (a) *Restrained elegance in a specialist morning paper,* the Financial Times

Page planning and typography 41

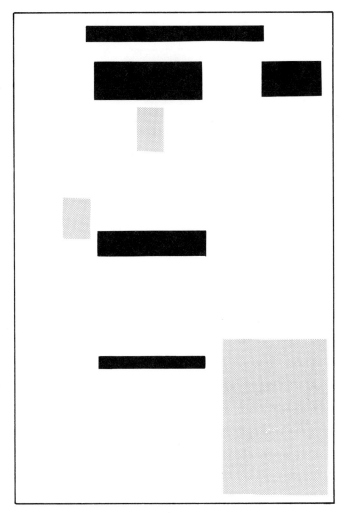

Figure 12 (b) *The page from the* Financial Times *seen as a pattern of highlights created by headlines, pictures and advertisements*

that does not have to compete – only to deliver the daily content of news and opinion on financial, economic and trade affairs that its readers expect. It settles for a restrained but elegant headline type of modest size and pictures that are functional and informative, and never many of them. The text is usually set single column and the layout is a low-key pattern of judiciously spaced out headlines, all in lower case type, seldom more than two columns wide, the features pages distinguished by a more pronounced use of white space between the items.

Figure 13 (a) *Artistic balance of display and reading text in a quality Sunday paper*, The Sunday Times

Page planning and typography 43

Figure 13 (b) *The Sunday Times front page seen as a pattern of highlights created by headlines, pictures, and advertisements*

Take a quality Sunday paper, *The Sunday Times* (Figure 13). Here, in a market not too socially dissimilar from *The Financial Times*, is quite a different approach. The display is consciously artistic with much use of horizontal rules, stories being taken across the page rather than down, the use of creative white space, mood pictures, working graphics, and drawn motifs to signal regular items. Though mainly in serif lower case, the typesizes vary a great deal, with the use of quotations or 'mood' labels as headlines and panelled in stories as breakers. The approach is an effective way of solving the problems of a large broadsheet newspaper with many pages and a high proportion of rather long features, a strong leisure and culture element, a relatively low news-of-the-day content and a sectional format.

Focal points

The four examples show how newspaper design is harnessed to do a particular job. What works for a popular tabloid would be as unsatisfactory for a town evening paper with many neighbourhood editions as it would be for *The Sunday Times*, while for *The Financial Times* to be projected as self-consciously artistic as a quality Sunday paper would mystify the readers used to its simple authority. Each approach has to be judged by how far it achieves what it sets out to do.

Yet if you draw boxes to represent the areas to which the eye is first drawn on each of the four pages (Figures 5(b), 7(b), 12(b), 13(b)), you will see that each page has one thing in common. It forms an assymetrical pattern in which the pictures and main headlines are the focal points.

These focal points themselves, through what they say to the reader, draw attention to the reading matter by which they are separated from each other. By moving from one focal point to another, normally from a larger one to a smaller one, the eye is guided round the page, alighting on the things that first attract and then taking in items of lesser prominence.

The location of the focal points depends upon the materials used and on the position and content of the advertising on the page. The pictures chosen can either be horizontal or vertical in subject, the stories long or short.

In placing the materials, account must first be taken of the advertising – what shape is it, does it contain mainly type or illustration? A picture should be kept clear of pictures in the advertisement, while headline type should not be alongside type if this can be avoided. The page examples given show how these problems are resolved.

It is because of this combination of circumstances under which pages are schemed that layouts vary in design even within one newspaper. In describing them as asymmetrical it should be added that a truly symmetrical layout (which is not found in newspapers) is virtually impossible because of the varieties of shape left for the editorial content by the advertising. The one thing common to all layouts, however, is (or should be) the creation of focal points as a principle of page design.

It can be argued that the focal points in the examples given above are bolder on the popular tabloid page than on, say, *The Financial Times* page. This is an indication of differences of approach to the readership market rather than of any fundamental difference in design philosophy. News is more pictorial in a national tabloid in the popular market, while the impact of picture and headline is less bold, for example, in *The Financial Times* than in some other papers. Yet the aim of the visual pattern is the same, each in its market.

Order

The order in which the headlines appear on the page, and also their size, have a message for the reader. They signify the relative importance of the text which they cover in relation to the other items on the page. Thus there is the main or *lead* story (Figure 11) which has the biggest headline mass and sometimes, though not necessarily, the longest text space. The second most important story is called the *half lead*. Then there are the intermediate *tops* (the term which used to signify top of the page is now applied to stories of intermediate length in any part of the page), and lastly a number of one- or two-paragraph stories, usually in the lower part of the page. These are called *fillers*. Sometimes they are schemed into the page; other times they are kept handy to fill spaces when bigger stories fall short for some reason. Even the fillers, however short, can cover an important news point and it is better to scheme them into the page to ensure that the news point is covered.

The reader who is first attracted to a page by a picture or a bold headline might, if time is short, scan only the main items. If there is time, the initial impact might result in the reader reading the whole page. A picture with its caption, for example, can act as a taster so that the reader is persuaded to read the story which it accompanies in order to find out the details. Some pictures, on the other hand, stand on their own with their own caption material separate from the rest of the text.

A way of looking at modern newspaper design is as a form of packaging by which the contents are commended to the reader. In this sense it signifies a breakaway from the old days when newspapers had columns of unrelieved small type with little or no illustration.

Type character

An examination of successive issues of the same newspaper will show that there is a consistent use of the same type ranges of perhaps one or two stock types with judiciously placed variants. It would be easy to work right through the type book in choosing type faces since most modern systems offer a wide range of types, yet this would produce a hotch-potch effect and make pleasing page design very difficult to achieve.

Yet the consistency, which is common to all newspapers, is not just a question of design. It is a way of giving a newspaper a particular visual character. This is a major item in its format by which, as we said above, it is differentiated from other newspapers.

Summing up the above points we can say that there are three principles of newspaper design:

1 To attract the eye of the potential reader by creating an attractive visual pattern.
2 To signpost the various items and signal, by the type sizes and placings, their relative importance on the page.
3 To give the newspaper a recognizable visual character by the consistent use of chosen types.

TYPOGRAPHY

It will be seen that page design is made up of four elements:

text   headlines   pictures   advertisements

Since the contents of the advertisements are already decided before the editorial material is placed, the designer – the person who schemes the page – can only note them. The emphasis on design is therefore on text, headlines and pictures. In dealing with the first two a knowledge of typography is vital.

There are two main type families by which the western, or Roman alphabet is rendered in print: the serif and the sanserif (or *sans*). The *serif*, of which there are two examples given in this chapter (Figure 14), is characterized by letters in which the strokes are of varying thickness with the ends finished off with a decorative flourish or tail which is called the serif. This style of lettering dates back to early Roman inscriptions and from it has sprung many type designs of which a number are still widely used by newspapers today.

The *sans* family has letters of mostly even strokes without the decorative flourishes or serifs – hence the term sanserif. Sans types, of which there is now a vast and ever increasing number of designs, appeared in modern times in response to the demand from advertising and newspapers for types with boldness rather than elegance. Figure 15 gives two examples in current use in newspapers.

In referring to types, the word *face* is used to mean the appearance of the type as it prints, while *fount* (pronounced font), means all the characters, i.e. letters, figures, punctuation marks, etc. of any one typeface in any one size.

Many of the hundreds of types in both families are used only for advertising and publicity purposes. Newspapers adopt a more conservative approach, choosing one or two readable stock types, which are then used in

Page planning and typography 47

abcdefghijklmnopqrstuvwxyz ß &?!£$ (.,;:)»«/⁻⁰'''°
ABCDEFGHIJKLMNOPQRSTUVWXYZ 1234567890

abcdefghijklmnopqrstuvwxyz æøß &?!£$ (.,;:)»«/⁻⁰'''°
ABCDEFGHIJKMNOPQRSTUVWXYZ ÆØ
1234567890

Figure 14 *Examples of serif type in popular use: Times New Roman (above) and Caslon Bold*

abcdefghijklmnopqrstuvwxyz ß &?!£$ (.,;:)»«/⁻⁰'''°
ABCDEFGHIJKLMNOPQRSTUVWXYZ 1234567890

abcdefghijklmnopqrstuvwxyz ß &?!£$ (.,;:)»«// ^'''
ABCDEFGHIJKLMNOPQRSTUVWXYZ 1234567890

Figure 15 *Examples of sanserif type in popular use: Gill Sans (above) and Grotesque (or Grot) No. 9*

various sizes and *weights* through the paper, with perhaps another face to give variety or special emphasis, or for features display. Of the two serif types illustrated, Times New Roman is the stock type of *The Times* and *The Sunday Times*, while Caslon Bold is a popular stock face with many provincial and some continental papers. Other popular serif faces are Century Bold, Bodoni and Cheltenham. Both the sans faces illustrated, Gills Sans and Grotesque, are popular tabloid paper types. Others widely used are Tempo, Gothic, Metro and Helvetica.

There is also a variation of the serif family referred to as the *slab serif* in which the serifs or tails are squared off, giving the type a characteristic chunky look. Examples are the Rockwell range, used in the *Daily Mail* features pages and in many American newspapers, and Playbill, a traditional type for theatre front-of-house bills (see Figure 16).

It is usual for a newspaper to opt consistently for a serif or a sans format and to stick to it. The sans format is more popular with tabloid papers, especially national dailies such as *The Sun* and the *Daily Mirror*, while a mainly serif format is to be found in the more traditional broadsheet papers such as *The Times* and the *Daily Telegraph*, with its use of Bodoni, and in many leading provincial papers.

abcdefghijklmnopqrstuvwxyz æøß &?!£$ („;:)» ã/⁻ᵉ¹⁻˙˙˙°
ABCDEFGHIJKLMNOPQRSTUVWXYZ ÆØ 1234567890

abcdefghijklmnopqrstuvwxyz ß &?!£$ („;:)» ã/⁻ᵉ¹⁻˙˙˙°
ABCDEFGHIJKLMNOPQRSTUVWXYZ 1234567890

Figure 16 *Slab serif types are more commonly found on features pages and in magazines: Rockwell (above) and Playbill*

Type use

Page design has to take account of the stock types in use in a newspaper, although within these there still exists plenty of scope for variety, and a type style can be evolved by utilizing the variants within a type range.

For example, types exist in *capital* letters and in *lower case*, and a noticeable modern trend is for newspapers, especially those using a stock serif face such as Bodoni or Century, to have all headlines in lower case. It is generally agreed by type experts that lower case type, with its more flowing contour and variety of stroke, is easier to read than capitals. Yet some popular tabloids using sans type have their main headlines in capitals on the ground that the boldness of sans capitals has more impact. Titling Gothic, used for the bigger headlines, is an example of a type that exists only in capitals.

Another variant offered by some types is that of an *italic* or sloping face (the slope being top right to bottom left) for use along with the roman, or upright face. Today, most of the italic faces in use are in the sans ranges, though italic type generally is becoming less favoured in newspapers.

Within a type range there also exist variants based on the thickness of the letter strokes. The standard *roman* letter is accompanied by a thicker, *bold* version and sometimes by a *light* and an *extra bold*. Century is an example of a type which exists in all four thickness variants.

Other variants are based on the width of the letter. Gothic has versions ranging from *extra condensed* and *medium condensed* to *bold* and *square*, while many types are designed with an *expanded* version.

Condensed faces are useful in the narrower columns of tabloid sized pages, with lower case the more favoured, especially in single column stories where the eye can move quickly down the column to take in the message. The condensed letters enable a bigger sized type to be used without limiting too much the number of characters (i.e. letters and figures) that can be accommodated on a line. Condensed faces are not much used across wider measures where they are less readable, especially in capitals. Expanded faces are noticeable in the more horizontal page layouts of the quality Sunday papers. The pages illustrated in this book demonstrate these points.

In the phototypesetters now used in computerized printing systems, the master film founts can be stretched or squeezed as needed to produce any degree of condensed or expanded forms at the stroke of a command key, or even tilted to give italic. Yet the need for precise control over the typography of a page has resulted in most newspapers programming their typesetters to deliver nominated type that corresponds to the founts designed by the typefounder rather than risk a hotch-potch of type variations and sizes on the page, arrived at through expediency.

Another useful attribute of photosetting is that the machine will deliver headline type tinted with a screen tint, or reversed as white on black (reverse video), which can be useful for display. More commonly, type is reversed on camera as needed.

Points to watch:

- It is best to use variants as such, while giving the main and greatest number of headlines to the stock type around which the pages are designed. Too many variants on a page can have the same confusing effect as too many different types. A mixture of widths caused by over-indulgent use of the photosetter would compound the offence.
- Where a different typeface is chosen for variety it is better for a paper with a serif type style to choose a sans face rather than another serif face that would clash, and vice versa. For instance, it will be seen that the two serif faces illustrated in this chapter, Times New Roman and Caslon Bold, do not match together. Nor do the Gill Sans and Grotesque. Where a different type range is being used on the same page as the stock range, it is better to use a type that is in opposition rather than one that is similar. Thus a bold sans face will give effective contrast against a lighter face with serifs, and Century Light an elegant touch to a mainly sans page.

Type sizes

The principal variation in type use is by size, the different sizes being used to denote the relative importance of each story on the page. Type (which can vary greatly in width, as we have seen) is measured by its height, using a 250-year-old system of *points* of which, as a guide, there are approximately 72 to the inch and 28 to the centimetre.

To simplify typesetting, typefounders in western and most other countries have, since the early eighteenth century, designed type to a series of set point sizes which have long been accepted as the basis of print design. These traditionally had such names as minion (7 pt), brevier (8 pt), long primer (10 pt) and pica (12 pt). Today they are simply referred to by their point sizes. The series runs:

5½, 6, 7, 8, 10, 12, 14, 18, 24, 30, 36, 42, 48, 60, 72, 84, 96, 120, 144

The sizes up to 14 pt are the standard text sizes. The reading text in most newspapers is in 7 pt or 8 pt. This is called their body type, while the opening paragraphs, or intros, are given 10 pt or 12 pt, or even 14 pt in the case of the page one lead or 'splash' story.

From 14 pt upwards are the headline sizes. Papers with a more traditional approach to type seldom go beyond 72 pt for their splash story, though headlines of up to 120 pt are common on the front and back pages of the popular daily tabloids.

It should be noted that while body type of up to 14 pt exists both in the serif and in the sans ranges, seriffed body type is more commonly used on the well tested theory that, as with wide lower case headlines, it is easier to read as the eye travels across the words.

There is one oddity to note about point sizes. They are derived from the metal base upon which typefounder's letters rose. This had to accommodate the 'ascenders' and 'descenders' which stuck above or below lower case letters such as f, k, p and q, as well as accommodating the capitals (Figure 17). This means that the constant depth of a letter, assuming there is neither ascender or descender, and which is called its 'x height', is smaller than the point size of its base.

With types designed with long ascenders and descenders, such as some of the Gothics and Tempos, the letters have a small x height compared to other types and look correspondingly smaller despite having comparable typesizes. Even capitals can look reduced, since they fill the space taken by the x height and the ascender, but not the descender. Only in the case of the various titling caps, used for big headlines, where there is no lower case, does the capital fill the full size of its base.

Choosing the typesizes is an important part of scheming a page since it gives the items the prominence the page executive intends, and also gives the page its visual balance. It is for this reason that phototypesetters, despite

Figure 17 *This example of Franklin Gothic lower case type has dotted lines showing the 'x' height of the letters in relation to the ascenders and descenders*

their capability of varying typesizes by intervals of a quarter or half a point right through the range, are usually programmed through their command keys to deliver type in standard nominated sizes. The facility remains to override the instruction to get a particular headline to fit, but if this facility is over-used the basis of the visualizing is endangered. It is also quicker to keyboard text and headlines in which sizes, and measures, are programmed in the machine.

Measure

As with the points system of typesizing, so with the setting width, or measure, have the old printing terms been retained in the computerized systems. Page widths and column widths are calculated in *ems*, an em being the width of a standard roman 12 pt lower case letter 'm'. An em is equivalent to a pica, which is the old name for type of 12 pt size since the 'm' it is derived from is as wide as it is deep. In US practice ems, in fact, are called picas. An 84-em page would be an 84-pica page; and a 9-em column a 9-pica column. In phototypesetting systems of American origin measures are usually commanded in picas and not ems.

Since an em equals 12 points it is useful to remember that 6 ems equals approximately 1 inch (25 mm), though to people familiar with printing, ems quickly become the norm in the calculation of setting.

Ens, based on the width of the standard 12 pt roman lower case letter 'n', are also used, particularly to denote measures that are indented to give white space either in front or on both sides, i.e. indented '1 en', or 'en each side'. The old printers' terms for ems and ens are muttons and nuts.

Newspaper pages have a standard number of columns. Most broadsheet pages have columns 9 or $9\frac{1}{2}$-ems wide, and tabloid pages 8-em or $8\frac{1}{2}$-em columns. In scheming pages, the column measure of both text and headlines is nominated as well as the typesize. It is the job of the subeditor, who carries out the actual editing of the stories, to see that these instructions are entered into the typesetting system.

In the case of standard single-column, two-column or three- column measures it is only necessary to nominate the measure in columns. This will be programmed setting in a photosetter. Any setting less or greater than single column or its multiples is called *bastard measure*, and here the instruction must be given in ems and ens so that it is set accurately. For instance, a special caption might be across $10\frac{1}{2}$ ems or an intro alongside a picture across $15\frac{1}{2}$ ems (See Figure 11).

DRAWING THE PAGE

Page layouts are usually drawn in pencil, since it enables changes and amendments to be made easily before make-up stage without them having to be redrawn. They are also duplicated so that copies can be given to the chief subeditor (for arranging the subediting), the editor or senior executive in charge of production, and the journalist in charge in the make-up area.

Where the art desk system is in use, pages are drawn in fine detail, including a representation of the pictures being used. Where there is no art bench, the page executive, or a nominated subeditor has to deal with the technicalities of layout, including giving instructions for reverse type, headline-picture compos, special setting formats, and the editing of pictures (see Chapter 4).

What is important is that the placing of stories and pictures should be accurate in size and length so that precise instruction can be given to subeditors, and so that pictures and text fit properly. It is little use, for instance, having a sophisticated computerized system which allows a precise length check on screen if the layout is so inaccurate that intros have to be reset and stories cut while the page is being made up.

Also important is the type balance. While the layout might seem to be a mass of instructions the typesizes and measures marked on it should reflect a considered pattern of headline and text with a balance of typesizes. It requires long practice as well as visualizing flair to achieve this. The pattern, as we have said, should take account of the shape and content of the advertising, and also the content and type balance of the page opposite as the reader opens the paper.

A problem can arise with tabloid-sized pages where a major item crosses over the pages. Since the pairing of pages at the final printing means that each page is made up with its printing pair (i.e. in a 36-page paper page 36 pairs with page 1, and page 35 with page 2 and so on) the making up of spreads has to be carefully checked so that the matching of headlines, borders and text levels are accurate, even though the two halves of a spread might be physically separated (see Chapter 12).

In addition to placing the stories and the pictures, page executives have a variety of means at their disposal for giving texture and design to a page. In choosing the types and measures the page executive will nominate special setting for some areas. This can be for emphasis or for visual effect. Many of these devices, such as the size of intros and the use of crossheads and of bold and italic body type, will be dictated by the paper's type style, although there is scope for innovation.

Intros are usually highlighted by being a size or two sizes up from the body setting, page leads being mostly in 12 pt and the intros for other tops in 10 pt. Most newspapers give the first word of the first paragraph in capitals, with

Page planning and typography 53

FIXED penalties applying to nearly 250 motoring offences come into force in England and Wales from Oct 1.

But chief constables are warning their officers to be "sensible" about issuing tickets and to continue to **W**hen the world's most famous recluse, Howard Hughes, died in April, 1976, on an aeroplane flying him from Acapulco to Houston for medical

Figure 18 *Drop letters – strictly outsize type markers at the beginning of a story – are now used mostly for decorative effect on features pages. The examples show a simple two-line drop on a news page 'top', and a monster stand-up drop from a colour magazine*

names capitalized in full where they begin a story. Some papers still carry 24 pt or 30 pt *drop letters*, which are initial big capital letters running on to a word in capitals of the intro type (Figure 18). This device is used extensively on features pages where more type 'decoration' is to be found. Where they extend above the line of the text they are called *stand-up drops*. Where intended, they should be instructed by the subeditor on copy as a two-line or three-line drop letter. For ease of setting in the phototypesetter they should be programmed in advance as a setting format.

A newspaper might use selected *bold* or *italic* paragraphs (never both) within a story to give 'colour' to the page or to highlight a point. They should be used sparingly to avoid a bitty effect, and should be marked up at the subbing stage so that they are set with the story, thus avoiding keyboarding them at page make-up and causing possible delay in the photosetting.

Crossheads, which are single or double lines of type, usually of 12 pt or 14 pt, are used to help the eye cope with long texts, and also to prevent greyness in page display. If set flush left they are called *sideheads*. In features, the type chosen for crossheads often complements the main headline type being used and is part of the display. On news pages, with their simpler type style, there are usually set types for crosshead, with capitals of a given size on lead stories and lower case on others. The newspaper pages illustrated in this book have examples.

While crossheads (and sideheads) are principally eye-breaks in the text they should, where possible, occur in natural breaks in the story and not interrupt thought sequences. Full lines should be avoided since they cut off text from text and halt the eye rather than ease the reading process. The type chosen should also be allowed at least 4 pt of white space above and below, whether in caps or in lower case, to keep the ascenders and descenders clear of the text.

There are two schools of thought about crossheads, arising out of their dual purpose. One is that the word or words should be a guide to the part of the text that follows. The other is that as they are simply eye breaks what they say matters little. Some quality papers, and also magazines, follow this through by using pieces of thick or decorative rule to break the text every few centimetres instead of type. Some newspapers dispense with crossheads altogether as a point of style.

If crossheads are used, their position should, where possible, be calculated in advance, especially in photoset systems, so that they can be marked up with the story text to avoid late setting, causing delays in page make-up. They will thus be delivered by the typesetter along with the text and headline.

The movement of crossheads on the page at make-up stage to achieve balance is limited by the fact that they must stand before the part of the story to which they refer. One way out of this problem, particularly with a long story, is to prepare in advance some small two-line *quotes* from the story which can be set in a special type bounded by 2 pt or 4 pt rules above and below, and used as breakers. These can be placed arbitrarily within the text to achieve the best visual effect, and can act as useful tasters to the story as the eye crosses the page.

Words in *bold capitals* are a useful way of emphasizing points in a story or introducing items in a list, as in shopping or market reports (Figure 19). Individual words or places can be picked out in bold capitals to give them emphasis. Another way to mark each item on a list is with a black *blob* of a

MEAT : Warming meals are back in favour, so choose slow-cooking cuts like stewing steak at £1·20 to £1·58 a lb, less than half the price of rump steak. Pork is another good choice, with loin chops £1·20 to £1·50 a lb, and boneless shoulder 98p to £1·38. English shoulder of lamb is reasonable between 78p and 98p a lb.
   FISH : Supplies are getting back to normal after the weekend gales, and best buys will be haddock fillets about £1·70 a lb.
   VEGETABLES : Carrots 10p to 18p a lb.; cabbage 15p to 20p.; **courgettes** 30p to 45p; green **peppers** 45p to 70p.

Figure 19 *The sort of news text that benefits from words highlighted in bold capitals*

bigger size than the body type, or with a black or open *square*, if available in the system. *Figures*, sometimes reversed as white on black, can also be used if they can be processed by the system with the text.

By-lines and *date-lines* are best set in a special type such as Metro medium caps or lower case at the top of a story to differentiate them from the intro.

While layouts are generally schemed within their columnar format, setting widths can be varied for emphasis or design purpose. A space can be divided into equal legs of a bastard width, as in the *Brighton Evening Argus* page, illustrated in Chapter 2. 'Colour' can be given by setting a whole story in bold lower case of the body type, where the stories around it are set in roman. Another device is to enclose a story in *panel* rules of 1 pt, 2 pt or 4 pt, or even in a shadow rule panel as in the page illustrated from *The Sun*, so that the actual setting is narrower than the standard width.

Photosetting systems enable time to be saved by being programmed to deliver setting formatted into panels of nominated sizes ready to be used in the page, with the added advantage that the story can be checked on the screen first to see that it fits.

Column rules of extra thickness can enclose stories that are linked, or rules between such stories can be omitted. Features pages often make a point of having white space instead of rules dividing stories.

Underscores can give a headline special prominence, but become a typographical cliché if overused. Headlines can also be used in *reverse video* – as white on black type or as black on a selected tint, as described above. This is a much used device in the popular tabloid papers, especially where a page is thin on pictures and needs highlights.

Special setting for such things as television programmes, race cards and football and election results needs devising with care to make it instantly readable, despite the weight of information it has to carry. Abbreviations and the use of bold and roman, caps and lower case should be consistent and meaningful, and several dummy runs should be tried before settling for a style that is informative and at the same time easy on the eye and not too space-taking.

Another great advantage of the computer is that such setting can be formatted on to one key, thus cutting out laborious and time-consuming marking up at the subbing stage.

The essence of all these devices, many of which are made easier through the use of photosetting, is that they should not be used too much or all together on a page. Discreetly chosen, they offer useful ways of improving the attractiveness and readability of a page design. They are all tools to be used by journalists who scheme pages.

4 PICTURES

The picture content of a page, as with the text and the headlines, represents the final stage of a process of planning and editing. In this case the result is that photographs of the right subject, size, shape and quality are presented to the reader. Let us examine more closely the role of pictures in a newspaper.

The notion that a good news picture is worth a column of words does not altogether hold true. There are few pictures that can stand without text whereas the text regularly stands without pictures, though there are occasions when it badly needs them.

There are cases of famous news pictures such as St Paul's Cathedral against a wartime background of flames, the shooting of President Kennedy, and the frigate Antelope (Figure 20) in the midst of a fireball in the Falklands War, that remain in the mind long after the words have faded. This is what really tells the story, the picture enthusiast will say. But, of course, on their own such pictures cannot – and do not. They have to rely on the context of words, on explanation, in the same way as tele-film relies on the commentary and the presenter's guidance. The better and more dramatic the picture the greater is the danger of the readers (or viewers) being misled by their own heightened reaction.

Shorn of their context of words and explanation, pictures can, in a real sense, lie or at least fail to deliver the truth. An example of this is the famous occasion on which a national newspaper reporting on the horrors of the Biafran Civil War, in Nigeria, carried a picture of an emaciated pot-bellied African child as a pictorial indictment of the perpetrators of the war. The child was not identified but it transpired that the picture was from another African country.

It achieved its object in bringing home its message to the reader, one might say. This is true but the picture was, nevertheless, lying. It had no connection with the Biafran War yet, placed as it was in the story, it misled the reader into believing that it had, for the reader was not given the

Figure 20 *A photograph that made news – the famous* Reuter *picture of the frigate HMS Antelope exploding into a fireball after being hit by an Exocet missile in San Carlos bay in the Falklands War*

necessary explanation of the picture's origin. If, because the intention seems to be worthy, this sort of practice is to be condoned, where does one draw the line in planting pictures to condition reader response? What right, critics of the press might ask, have editors to manipulate pictures in this way?

Yet, notwithstanding the possibility of misuse, pictures are a vital part of news coverage and a newspaper would be less informative, less complete and less attractive without them. They not only supplement the text, they enhance and extend it by highlighting and pressing upon the reader important parts of it and make it easier for the reader to build up a picture of what he or she is reading about. This is true whether the picture is of an occasion or news scene or just a 'head shot' of someone involved in a story.

News pictures in a newspaper have a permanency and graphic effect denied to television film which has a fragile grip on the senses and a low level of recall. Being in black and white can increase the drama and tension in a picture compared with colour, as anyone who has studied colour in newspapers is aware. It probably has something to do with the basic unreality of the colour range that systems give compared with the nuances of tone and texture in real life. Black and white plays on the imagination and reconstructs reality in the mind whereas colour blandly invites one to accept its artifice, although it has an important visual impact.

If the reader is to be correctly informed, however, and not to fall unwittingly into error, the relevance of the picture to the story must be made clear in the text and, in particular, in the caption that goes with the picture. No picture should be published without explanation or clear identification. Even a so-called 'mood shot' accompanying a feature can raise doubts or misconceptions in the reader's mind if it is left uncaptioned.

Pictures also serve another function in the modern newspaper. They form distinctive eye-catching areas to which the reader's attention is directed when he or she turns to the page. The position of pictures in relation to the contents of a page is thus carefully chosen as part of the balanced visual pattern by which the newspaper's content is projected. We can say, in short, that a picture performs two functions:

1   It illustrates the text.
2   It forms an element in the page design.

PICTURE SOURCES

The taking, collecting and providing of pictures on a newspaper is organized on similar lines to the newsroom operation. It revolves around the picture desk, or picture department, which works in close conjunction with the newsroom and the picture library. Directing this operation is the picture editor (on a bigger paper) or chief photographer.

The picture editor keeps a schedule of jobs in which pictures are wanted, or are expected, and ensures that photographers are briefed about the paper's requirements and the various deadlines. The photographers might work alongside reporters on a job or work independently, depending on time and the sort of job. Alternatively, if the use of a staff photographer is not possible or appropriate, then it is the picture editor's job to provide pictures from other sources.

Staff photographers

Staff photographers, working from head office or branch offices, or sometimes operating independently in a district, are the main sources of exclusive pictures and have the advantage that they can be more easily deployed alongside reporters on jobs where on-the-spot pictures are wanted.

Press photographers are journalists by training and definition and their approach and job briefing are in line with that for reporters. They share the same deadlines. Their staff numbers vary from four or five on a small paper to perhaps twenty to thirty on a national daily. The newspaper's own darkroom prints their work alongside the developing and printing of any other film required in the course of editorial production.

Staff photographers are the most controllable source of live pictures since detailed briefings can be given of the sort of pictures needed and special things to be looked for. Also, transmitting, transportation and processing facilities can be provided so that the pictures are ready on time.

Staff pictures become the newspaper's copyright, and syndication or reproduction fees are earned if they are used in other publications.

Staff photographers, even when working alongside reporters, are expected to be responsible for their own caption material and fact and name checking.

Freelance photographers

Freelance photographers provide some of the picture input for newspapers but, except in the case of high-fee exclusive work to which they hold copyright, or specialist areas such as glamour, their contribution is usually by daily, weekly or holiday relief arrangements or for a one-off job or series of jobs. In some small towns or provincial areas local newspaper photographers might work on a regular freelance basis for bigger papers by being accredited in the same way as local correspondents. This might involve a small annual retainer fee plus payment for each picture used, or for those ordered even when not used.

In the more specialist areas such as glamour, sport, animal and technical photography there is a good income to be earned by freelances whose skill becomes known and to whom newspapers, and more especially magazines, turn at times of need. Their picture credit lines become as well-known as the by-lines of big name writers.

Picture agencies

Picture agencies provide a great range of pictures, both news and for stock, especially from overseas sources, often on a contractual basis, to subscribing newspapers. Such 'service' pictures, though lacking exclusiveness, usefully fill gaps in coverage and can sometimes provide the only picture available. Some agencies have specialities such as sport or political or celebrity portraiture. There is usually a reproduction fee for each occasion of use, unless an agency is prepared to part with British or other reproduction rights of a picture at a higher fee in cases where, for syndication purposes, a newspaper wants to control the use of the picture.

Rota pictures

Agencies are sometimes relied upon on special occasions where only limited picture coverage of an event is allowed on a rota basis. Royal or certain political occasions, or even celebrity 'photo calls', where a large influx of press photographers is not convenient, are sometimes on a rota. The rota is organized by a committee drawn from picture editors representing the various newspaper publishing associations and agencies and has a permanent secretary who keeps a list of those entitled to camera facilities. From this list, newspapers and agencies take it in turn to cover rota jobs, on which anything from one to perhaps six or seven press photographers are allowed. The resulting pictures must then be made available for use to the newspapers subscribing to the committee's arrangements. The advantage of the system is that all newspapers who are interested have the chance of a picture. The disadvantage is that an exclusive picture is not possible.

Collected pictures

In some reporting jobs, and also with features, picture material might be supplied by the subject or organization concerned, especially in cases where the use of a photographer is inappropriate or impossible. Permission is needed for the use of such material and any possible rights investigated.

Hand-out pictures

A good deal of the material used with TV programmes and showbiz stories and columns is supplied free to newspapers on a regular basis from filmmaking and programme companies. Free pictures often accompany other hand-out material where the aim is to seek publicity through use in a newspaper.

Picture library

All newspapers keep indexed files of news pictures and pictures of people who have been, or are likely to be, in the news (such as MPs and sports people) going back as many as twenty or thirty years. The library is used where 'head shots' or recent likenesses are needed with stories or when an event that has been photographed comes back into the news, or for the purposes of 'flashbacks' to similar events in the past. The files are also a useful check source for identification and personal details, as with the news cuttings library.

From these various sources flow the input of pictures to a newspaper's picture desk where they are sorted, checked and identified and then submitted to the executives in charge of planning the pages who select from them the illustrations for the day's editions.

PHOTO BRIEFING

The picture editor is responsible for briefing photographers, and any other sources, about a newspaper's picture requirements. Usually he or she has been a photographer (it does help) and so is aware of the factors in photojournalism. The picture editor will know, from the daily news schedule, and the editor's conference, the nature of each picture job and if there is a need for a specific sort or shape of picture or the inclusion of certain wanted detail.

Technology has leapt ahead in press photography and has helped to make some formerly difficult assignments a little easier. Long heavy telephoto lenses for which the help of an assistant was needed are a thing of the past. The use of mirror lenses for press work has effectively halved the length and weight of the telephoto unit required by enhancing the magnification of the image. They also increase the amount of light reaching the film, making for a better quality picture.

The focal length of telephoto lenses can be doubled by the attachment of a powerful supplementary lens, while the use of an electronic booster between the lens and the film, by picking up the light at the red end of the spectrum, makes it possible to take a picture in almost total darkness.

With new advances in miniaturized portable transmitters, films can be developed on location and wired, even in colour, so that they reach a newspaper office in a matter of minutes. New 'negative' transmitters will wire directly from a negative, reversing the image so that the picture is received in positive. It is possible with another device to take good prints from video film which make acceptable pictures for newspaper use. Though new devices take time to reach general use, developments in press photography are towards faster, more immediate picture transmission, which helps in the planning of pages.

The modern tendency in news photography is for greater use of telephoto lenses of up to 400 mm, or occasionally up to 1000 mm, in order to concentrate on people and actualities, and there is less scope for the sort of composed mood shots beloved of Robert Capa, Henri Cartier-Bresson and, more recently, Don McCullin. The decline and closure of news picture magazines, and the limited news scope in colour supplements plus the almost exclusive use of 35 mm cameras and clusters of shots are perhaps responsible for this. One could add that in atmosphere and mood, television is not easy to beat.

With telephoto lenses comes the opportunity of exploring the unusual angle which can enliven a page – what Harold Evans calls 'the yawn in the crowded political meeting rather than the candidate in the centre of the waving crowd'.

In a news picture the content is the prime factor, with animation rating high and a good likeness vital. People and actualities are what hold the attention of the majority of readers when they turn to a page. This is nowhere more important than in the scenes and faces photographed for a local paper.

News pictures are usually of two main sorts. There is the action shot which helps tell the story, and there is the sort that shows people, which might simply inform. The first is the more prized and is the harder to catch since there might only be one chance. The second can be set up simply by the photographer asking people to take up positions. It is here that the danger of a cliché picture arises, especially if the photographer is overworked or rushing from job to job. Yet ingenuity can produce an out-of-the-ordinary composition.

In big events such as state funerals, Trooping the Colour and festivals, pre-arranged shooting positions, and the lack of mobility, make timing of the essence in getting the wanted picture. In indoor work the effect of the flash on the subject is the perennial hazard plus the problems of shadow and small room perspective. It is here that a wide-angle lens is useful.

The skill of the picture editor lies in taking account of the various hazards when briefing for special requirements.

Photographers usually develop and contact-print their own film and it is from the contacts that the picture editor, looking for sharp detail, selects and masks the frames for enlargement so that a selection of pictures is available from each job for the page planning stage.

Here again, technology has stepped in by devising systems that digitize and store incoming pictures on magnetic discs from where they can be recalled on a monitor screen and examined for choice. Thus, under this system, only the pictures to be used need be printed.

CHOOSING PICTURES

Even without stock picture sources, newspapers have many more pictures available each day than they can possibly use and a selection process takes place similar to copy-tasting. The following points are taken into account in choosing a picture:

Content

The relevance of the picture content to the story it illustrates. Does it show the right people and scene? Is it a good likeness? Is the action shown and the background of the picture right? Does it properly complement the story and extend its content?

Composition

Do the grouping and position of the people in relation to the elements of the picture form a pleasing 'shape'? Is it the most eye-catching of the pictures that have the required subject content?

Balance

Will the content of the picture balance well with the rest of the page? For instance, 'head shots' should not look out of the page or away from the story they illustrate. Also, clashes with other picture content on the page, including the advertisements, should be avoided. A picture of an ocean liner will look superfluous on a page containing an advertisement also showing a picture of a liner. Likewise, a picture showing the heads of a man and woman would lose effect against an advertisement containing the same sort of picture.

Tone

Has the picture good tonal values? Is there sufficient contrast between the dark and light tones so that it will reproduce adequately, bearing in mind that most newspapers print in black and white and on newsprint which is nowhere near as kind to detail as glossy bromide or gravure? A lack of dark or light tones can result in a grey effect which blurs detail on printing, although it might seem adequate on the original print.

A temptation in choosing a stock *picture* is to condition the reader by selecting one that shows a person in a particular light – the grim face of a trade union leader, the smug or triumphant expressions of politicians, or pictures that send up well known people by showing them as they looked when caught by the camera in an unguarded moment.

The choice might seem fitting for the accompanying story but there is a line beyond which the display of foibles and character can become bad taste or propaganda, or even libel if a person is really made to look stupid – as lawyers are willing to demonstrate.

THE DESIGN FUNCTION

The needs of page planning can influence the choice of a picture and its shape and size on the page. Since pictures, as we have said, form part of the visual pattern of the page under modern design techniques, it is usually desirable to have one main illustration on the page with perhaps a number of small ones, or maybe no others at all. The choice of the main picture has to be made at the start, for it governs the page design to a large extent.

The best picture for visual effect might not necessarily go with the main story. In fact it might be with one of the minor items or be covered simply by a self-contained caption and be chosen for its picture value in the absence of any usable picture with the main news content. The visual balance of the page is thus seen at work as a factor in picture choice. The main picture is a focal point in this balance and its size and shape is integral to the structure of the page, its position being influenced by the relative visual mass of advertisements, main headlines and text, all of which go to make the design. The illustrated examples of pages in this book, which are discussed later, show how this factor works.

Taking this explanation a stage further we can say that the inclusion of a story on a particular page can be because it provides a picture which can help give the page visual balance. It is not usual to have two stories with good picture coverage on the same page, leaving none for another page. An exception to this would be on a page on which the intention is to present news in pictures. Only where the editorial space left by the advertisements is very small would the page executive perhaps not bother with pictures.

The quest for visual effect can become an obsession in the case of some tabloid newspapers and picture display, in the end, can starve the pages of reading matter if pushed too far. The idea, on the average newspaper, is to strike a balance between pictures and text so that both functions performed by pictures are properly utilized and a newspaper, while making itself attractive, can get on with its job of being a newspaper.

USES OF PICTURES

Where a certain type of picture is expected, or is going to be used almost for certain, it is usual for the picture editor to brief a photographer with such advice as 'a strong double of this' or 'go for a wide shot – it's the heads that matter'.

The worldly cameraman, since he has to shoot off his 36-frame film, will probably go for a variety of shapes and angles for good measure, so that there is a choice if the advertising department has left an awkwardly shaped editorial space to fill, or a story is moved to a different page.

Nevertheless, the picture finally chosen will determine by its composition, and what it is required to show, the shape it will take on the page. In fact a picture considered to be worth a good space will be given a page where this is possible, and its shape and size will govern the disposition of stories and headlines around it. It is a fixed point in the page design to which text and headlines adapt. It still has to perform its function of informing the reader, even when it is part of a picture sequence, since it is an ingredient of the day's input of news.

If a number of pictures are to be used together on a news page, a *sequence* is better than a *montage* (the coupling of pictures into a display pattern, with complicated cutting in). This is not only a troublesome and time-taking device; it can also make the page look contrived.

Paste-up layout and photosetting make a great variety of display ideas possible but on news pages there is nothing faster or more urgent than a well-cropped, good-sized perfectly toned rectangular picture. Yet at various times the classic rectangle has been threatened by the *cut-out* in which part of the picture is highlighted by being cut away to reveal the outline of a head or building or other salient feature.

Under the hot metal system of blockmaking the engraved plate reproduced the cut-out outline of the print from which it was derived. With paste-up page layouts derived from photoset type, the use of cut-outs is easier since the photograph, having been edited and printed to the correct size, is simply trimmed to the required outline and attached to the page.

Reproduction by the half-tone process, in which the tones of a picture are rendered by means of tiny printed dots of varying density, still applies in computerized systems both in colour and black and white. Pictures are still printed, in effect, through a fine screen, as they were in the hot metal process, though with the use of polymer printing plates on web-offset presses a finer screen is possible, thus giving greater clarity of reproduction.

Screen make-up, though now beginning to be used by some newspapers, depends on the graphics facilities that are built into the system, and has its limitations. Pictures, which are at no disadvantage in an electronic system, here tend to form parts of modular design patterns, though these need not preclude cut-outs related to decorative bastard setting around them. Display advertisements are good examples of the uses that can be made of cut out pictures in a decorative way with the use of photosetting and screen make-up.

Cut-outs are used mostly on features pages. They look over-decorative on news pages and detract from the urgency. Likewise, the use of pictures as part of a *compo* (or composite illustration) along with headline type, sometimes reversed as a *WOB*, or white on black, or printed across the picture, is a fussy, though eye-catching device that is kept mainly for features display, of which examples are given later in this book.

On news and features pages a simple but significant picture subject might be used as a small *bleach-out* in which the detail is over-exposed in the developer until it loses its intermediate tones and adopts the hard black and white of a line engraving. This technique is used where a motif, or series of repeating motifs, is wanted to flag a long-running story – such as pit-head gear in a pit strike or the outline of the Kremlin for a Russian story. Bleach-outs, if chosen wisely from a sharp original, can have the drama a drawn motif lacks.

PICTURES AT WORK

The four news pages reproduced in Figures 5 to 8, and referred to in Chapters 2 and 3, demonstrate the uses of pictures in newspaper design. The tabloid size *Brighton Evening Argus* news page has a good example of an animated local news picture of the mayor switching on the Christmas lights. Here the two subjects are responding well to a merry occasion in what might have been otherwise a dull conventional pose. The reproduction of the letter in the top story on the page shows how a hard ragged edge drawn round the relevant part of the words can make a letter stand out as a picture. Applying this edge to a bromide of the original is the retoucher's job.

The broadsheet *Western Mail* news page uses two Welsh-interest pictures of twins (from a Multiple Birth Association conference) to create a panelled-in focal point below the page lead, in the centre right of the page. Despite having only one small advertisement on the page, the one picture display is sufficient to carry the headlines and text of the ten stories grouped around it, and is a good example of how a big page can be given structure by the use of pictures. The image creation of the pictures is also good, with the big one being cropped for action and the smaller for prettiness. The remaining focal points on the page are created by the balanced use of bold and light Century type in a wholly lower case format.

The Sun page settles for a simple deep two-column of the family in the lead story, fairly closely cropped as a foil to the rather jazzed-up advertisement. A small head shot at the top of the page creates a breaker between two lots of body types. The *Daily Telegraph* page is an example of a traditional broadsheet layout complete with five-column political cartoon as the main focal point, and with no half-tone at all.

In all the news page examples shown, the editing of the pictures is conventional with no 'artistic' cropping or straining after visual effects. The features pages analysed in Chapters 14 and 15 demonstrate quite a different approach to editing and placing. The *Liverpool Echo* double spread and *The Guardian* page illustrated there, both show an imaginative use of pictures as an integral part of the display. They are artistically cropped and are placed to give the maximum visual impact, interfacing with the type to create page patterns.

PICTURE EDITING

Pictures have to be edited to prepare them for reproduction just as text is edited. This work is done not by the picture editor (who, as we have seen, is the executive in charge of the taking and commissioning of pictures) but by the page executive or layout artist. The following are the sequences:

1 *Cropping* Selecting and marking off the part of the picture to be reproduced.
2 *Scaling* (or sizing) Calculating and marking up the size and shape of reproduction needed.
3 *Retouching* Improving detail and removing defects from the picture by brush and pen. This last operation is carried out by a retouching artist who sometimes works in the photographic department.

Cropping

Very few photographs are printed exactly as they are taken. A picture might include detail which is irrelevant to the story it accompanies. This might be because of difficulties in taking the picture. Perhaps it includes people who have nothing to do with the story, or maybe the page visualizer wants to bring up a part of the picture, or exclude part, for reasons of news value or page balance, or even for legal reasons. The procedure to achieve this is called cropping and it is quite simple.

The selected print (not the negative) is turned back to front on to a light box or against a light source, and the area to be excluded is marked off in pencil on the back. The lines thus not only enclose the image required but nominate its shape as it will appear on the page (Figures 21 and 22). Cropping can also be done by grease pencil on the front of the picture (but is messy and can damage the emulsion) or can be done by masking the outline of the image in tracing paper on the front (this can also leave marks on the emulsion unless done carefully). Pencil marks on the back can be easily rubbed out if the picture is to be used again.

The shape resulting from cropping (anything other than a rectangle is not normally used or feasible; even a cut-out, or oval, would be developed from a basic rectangle) is governed partly by the natural composition of the picture, partly by the intentions of the page visualizer, and partly by the space available on the page. The picture's composition predominates,

Figure 21 News of the World *cameraman Brian Thomas captures the moment of triumph in a big boxing match – the photograph as first developed*

however, and it is not usual to force a picture of naturally horizontal shape into a vertical. If its natural shape is unsuitable then it is either the wrong picture or it is on the wrong page.

Where a system has a full graphics facility, it is possible to crop pictures electronically on a screen where they have been stored in the computer, but it is more satisfactory to crop the picture as needed before entering it into a system for screen page make-up.

The cropping of pictures can arouse violent disagreement and much pontificating by experts who think it should have been done differently or sometimes not at all. A good picture, it is said, is its own advocate, and (insisted distinguished photographer Henri Cartier-Bresson) should require no cropping. This can be true if the picture is being used for what it is and not for what it is required to do, as for example a new royal portrait. Not many pictures fall into this category for in a newspaper pictures are subservient, though complementary, to the purpose of covering the news and informing the reader.

A newspaper is not just a vehicle for the camera. Also the requirements of big picture display, however attractive to the eye and justified by picture quality, have to be balanced against demands on space by the news coverage to which the paper is committed by its style and market. It is worth remembering that it is in detail of information that the newspaper scores

Pictures 69

Figure 22 *The picture as it was cropped and used to show the two fighters in close-up*

over picture-dominated television, both in editorial content and in advertising. The eclipse of newspapers, and especially magazines, that have sold primarily on pictures, is on record.

An experienced cameraman knows when he has got a good picture in terms of light, composition and content. The way the picture is cropped and used on the page, however, and the significance with which it is imbued, depends also on the words of the text and the requirements of page planning – its functional role as an illustration to the story and its structural role as a focal point in the design. The aim is to get the best possible image out of the picture that will make these points, and its success is measured by how far this aim is achieved and not by any abstract standard of what constitutes good camera work.

The cropper seeks to avoid wasted grey areas being reproduced, enhancing the composition while at the same time emphasizing required detail.

Thus, in a picture showing a group of houses, one in particular might be selected because it is the one with which the story is concerned. A person taken from a group picture is cropped from the waist upwards because that, in the view of the visualizer, is what the story needs. Stock pictures are often used for this purpose. An agency news picture might appear in rival newspapers cropped in different ways because of different ideas in projecting a story.

Cropping should be creative. It is a method of editing a picture for emphasis. Bad cropping occurs:

1. When the position or attitude of the subject person becomes distorted or meaningless because of the exclusion of necessary detail.
2. When the cropped area is too small, causing a dispersal of detail on enlargement and printing.
3. When lines and perspective are thrown out by 'tilting' in an effort to get a wanted shape.

Scaling (or sizing)

The scaling or sizing of a picture, whatever the reproduction process, is really a calculation of its depth since the width is chosen at the outset as a starting point in the page design, whether it be single-column, double-column, three-column, or any other width in between or greater. Here (Figure 23) is the method for calculating a picture's depth:

Draw a pencil line diagonally on the back of the print from a top corner to a bottom corner of the cropped rectangle. Measure across the required width, whether it be two-column, three-column or whatever, until the measurement intersects the diagonal. Then measure to the foot of the cropped area. This gives the depth of the picture as it will appear. The depth increases or decreases in fixed proportion along the projection of the diagonal.

The same method of calculation can be used if the depth of the picture is decided first and the width has to be calculated. The measurement is done in reverse by intersecting the diagonal by a line drawn up from the bottom of the cropped area and then measuring across.

A reading can also be obtained by using a 'magic wheel' or proportional scale which has two calibrated discs revolving on a common axis which can be turned to give a reading in a window. This method is useful on magazines

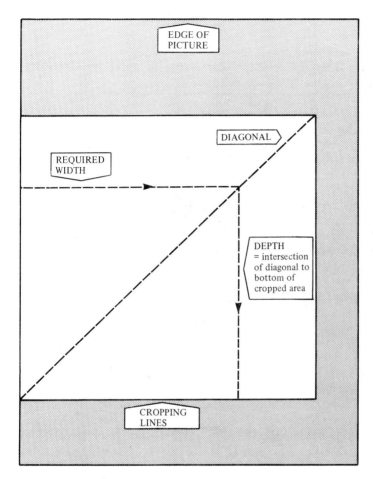

Figure 23 *Scaling, or sizing, a picture by the diagonal method*

which deal with small positive transparencies which are otherwise hard to size. A slide rule can also be used.

Retouching

Inevitably there are problem pictures. An example is where a picture of inferior quality – or sometimes of quite poor quality – is used because it is the only one available and is vital to the story. It could be a picture of a murder victim or a missing person. Another example is where, despite cropping, a picture has unwanted detail such as someone's head which has nothing to do with the story, or an interior in which the heavy tones of wallpaper clash with a person's features. This is where the skills of the retouching artist come in.

The use of an air brush can alleviate harsh grainy texture, while a fine sable brush can bring up faint detail, provided the detail is there to start with. If this is done discreetly a poor picture can sometimes be made into an acceptable illustration. If overdone it can look like the proverbial 'oil painting'. It is a failure if retouching lines are visible on the printed picture.

Likewise, unwanted background can be removed, provided this does not damage the content or significance of the picture or turn it into a nonsense or, worse still, mislead the reader. In other words, the picture must not be made to look something that it is not. The removed portion must be genuinely extraneous and its absence no loss.

There is a question of journalistic ethics at stake here and the Press Council in Britain has ruled that retouching should not go beyond the bounds of enhancing and clarifying a picture to the reader to a point where it becomes misrepresentative. At this point it is possible for the law to step in, for 'doctoring' of detail can perpetrate a libel.

It is worth noting here that retouching is considered in law to be an editorial function despite the fact that the retoucher, in some newspaper offices, is not classified as a journalist and works in the newspaper's photographic department. The editor is legally responsible for what the picture shows in the same way that he or she is responsible for what the text says. In fact, retouching, other than routine strengthening of lines and lightening of backgrounds, should be done only on editorial instructions.

In some cases pictures, at editorial request, might be *reversed* in printing when needed for the balance of the page, i.e. so that people are looking the right way for good visual effect. This is a dangerous practice unless the detail is carefully studied. Some readers are good at spotting coats buttoned the wrong way, wedding rings on the wrong hand, or hair partings on different sides in different papers. More extreme results are right-hand drive cars turned into left-hand drive cars and lettering the wrong way round in background detail and car numbers. Some newspapers understandably ban the practice.

Colour work

The introduction of high speed web offset presses able to cope with longer print runs and tougher use has made possible, and cheaper, the expansion of colour on editorial pages. Whether this is successful in terms of reader appeal, or even desirable, is a continuing debate in newspaper offices. Certainly there is now plenty of equipment available to help achieve the highest quality effects from colour transparencies used in editorial pages and in advertisements.

Electronic colour scanners can enlarge transparencies up to ten times, enabling them to be cropped and scaled precisely and corrected for tone and quality. Separations are produced by means of microprocessor controls and from these, plates are made for the four-colour printing process used in run-of-press colour. The tones are rendered on the printed page by fine screened dots of colour all but invisible to the naked eye. They appear smooth, of course, on the web offset printing plate, unlike the raised dots which control the tonal density in half-tone letterpress reproduction.

Nevertheless, page changes and actual printing times are slower than with black and white because of the need to have extra plates printing in register to achieve the correct colour balance. An advantage is that, unlike pre-printed colour, the run-of-press four-colour process used with web offset does not need special newsprint.

It is also essential, in choosing colour transparencies, to apply the same rules at the planning stage as with with black and white prints if a transparency is to produce a satisfactory effect and look anything more than just a splash of colour on the page.

ETHICS AND TASTE

Retouching, as we have seen, can in extreme cases perpetrate a libel. There are also other matters of taste and legality which weigh with newspaper executives who choose pictures for publication. Violence and obscenity, for instance, are common dangers. There are pictures available, often of startling news potential, of death, injuries, the results of torture and starvation or shots connected with vice and drugs, which an editor might shrink from printing to avoid upsetting the readers. The news value of a 'shock' picture has to be weighed against the revulsion it might cause at the breakfast table. Where such pictures are used there has to be a sound reason of public good, and the decision to print is not an easy one.

The objection against pictures taken as a result of intrusion into people's homes and lives has to be considered. Such pictures can cause distress and bring an editor into collision with the Press Council, which regards intrusion as a serious breach of good practice which could force the law down on editors who are not prepared to submit voluntarily to the Press Council's directives.

Complaints can occur in diverse ways. A British national paper once had a lawsuit on its hands over a wedding picture in which the press photographer, in his zeal to photograph an heirloom garter worn by the bride, produced a picture which the girl's family claimed, when published, was degrading and obscene.

Pictures, more so than the text, can arouse deep emotions, and there is a responsibility upon those choosing them to consider this effect and not to go for unjustified sensationalism.

GRAPHICS

Graphics is now taken as a general term meaning any drawn visual aids used in newspaper and magazine display as opposed to photographs. It was originally limited to the use of design symbols to explain statistical material – the little people and half people signifying voting intentions, petrol consumption, budget benefits, etc. – and examples of stylized 'idiot' graphics have been available for years in Letraset and other stick-on-type systems.

Editors of old, hooked on the actuality of photographs, were not keen on drawn material and few newspapers retained an artist on the spot who could provide it quickly to order. Many of the wartime front-line maps were issued through agencies, taken from hand-outs or even traced from office atlases and filled in by a retoucher, or, in some cases, provided on the side by the paper's political cartoonist, if it had one.

The increased use of trained layout artists in recent years on the bigger newspapers as well as on magazines has led to a boom in in-house graphics as illustration, especially in the national Sunday papers. There are occasions when a drawing can get closer to a story's actuality than a photograph. Tabbed diagrams can show methods used by an escaping convict, the infiltration of terrorists, or the opium trail from Thailand, in a way denied to the camera and can make the words more real and understandable to the reader.

Royal family trees can come alive out of the routine researches of genealogists, the range and capacity of inter-continental ballistic missiles and Star Wars systems can be demonstrated, and football moves and golf strokes shown in detail.

Graphics can combine with half-tone by juxtaposing tabbed identification panels alongside important group scenes. It can demonstrate with arrowed lines on a photograph the path of a bullet or the route of a careering vehicle.

Maps continue to be an important contribution and the pictorial ones of the Falklands War showed how far technique had advanced since the dotted salients of the Second World War front-line maps. Maps can not only convey big areas but also the immediate geographical layout of a new situation and can be a functional aid to the text. They are by no means 'slow' alongside pictures and are worth using a lot more than they are used at present. Nothing gets in the way of their message (Figure 24).

As with pictures, diagrammatic graphics require a justification. They must materially extend the text and should be used sparingly and significantly and

WORLD NEWS

The two faces of Russia: honesty on the liner tragedy, but still the old chicanery on spying

Disaster strikes top workers on trip of a lifetime

by Louise Branson
Moscow

THE ORCHESTRA play'd as the Admiral Nakhimov went down last week, but few of the victims heard it. Most of the 400 people who lost their lives were jammed into narrow corridors, unable to reach the upper decks in the few minutes before the passenger liner sank after being rammed by a cargo ship whose crew may have been drunk.

Unlike the Titanic, there were no wealthy or famous names among the 1,234 people — 346 of them crew — on board the battered 61-year-old liner as it sank in the worst catastrophe in the Soviet Union's maritime history. For many passengers, the Black Sea cruise from Odessa to Batumi, near the Turkish border, was a reward for outstanding work at the factories and offices. Their trade unions had paid more than 60% of their fares which ranged from £150 to £30C. In the hours before the disaster the holidaymakers had been sight-seeing in Novorossiysk. The sun had been beating down fiercely on them all day as they trailed around the town.

They visited the fortress built in 1838 and saw the eternal flame to the memory of the dead of the second world war. They also heard Shostakovich's requiem which was composed for Novorossiysk's war heroes. It was hardly surprising that by the time the ship set sail in the evening, on a calm sea and at a steady 10 knots, the passengers were tired and hundreds took the fatal decision to go straight to their cabins and to bed.

Others, like the young honeymoon couple Alyona and Yuri Pavlikovsky, were wide awake in upper-level cabins or had chosen to stay on the central deck to listen to the ship's orchestra. Although they did not know it then, staying awake saved their lives.

Just 45 minutes after setting sail, the liner was rocked by a collision. A cargo ship, the Pyotr Vasev, loaded with 41,000 tons of grain, ploughed into the liner's forward starboard side, rip-

① 10.30PM Sunday
Liner sets sail

② 11.00PM (Approx)
Liner captain warns cargo ship of danger. It replies 'Don't worry'

Going down: despite being warned that it was on a collision course, the Pyotr Vasev ploughed into the ageing liner, the Admiral Nakhimov, as passengers went to bed

③ 11.08PM (Approx)
Liner repeats warning

④ 11.15PM
Ships collide

⑤ 11.22PM
Liner sinks

ping it apart. In contrast to the Titanic's leisurely descent, the Admiral Nakhimov sank in less than eight minutes. There were plenty of lifeboats but there was no time to get to them.

Of the 1,234 people on board the liner, 836 were saved and 29 of them are still in hospital, most suffering from pneumonia. So far, 116 bodies have been recovered and 282 people are still unaccounted for.

The tragedy has only one possible explanation: human negligence, possibly caused by drunkenness. It is a heavy blow for the Soviet Union, coming on the heels of the official report into the Chernobyl nuclear accident which also put the blame on human error.

As a result of the preliminary investigation, he Geidar Aliev, a P member, both ca Vadim Mankov, of t and Viktor Tkachenk cargo ship have b rested and accused o to obey safety regula convicted they are l face heavy penalties

There is confusic which ship was a Under Russian law ai ship must give wa international law sa: are obliged to give wa vessel approaching cargo ship was, in

and navigators frequently were drunk during voyages.
The death toll would have been far higher if the ships had collided further from the shore. The first rescue teams were able to cover the seven miles to the collision within 11 minutes. In less than an hour, 64 rescue boats were plucking survivors from the warm sea. Ten helicopters from the nearby resort of Sochi hovered overhead to light the midnight skies and guide the rescue work.

In the glare of the helicopters' searchlights, the sea appeared as a film of oil and paint dotted with bodies, barrels, debris and hundreds of people. Some were clinging to life rafts, some had lifejackets, but most were swimming or treading water.

Women and children were the first we tried to rescue," said Balikhan Khazarov, a sailor. "Our fellows saved three girls who were terrified, all covered with oil. They seemed not to even realise what was happening to them."

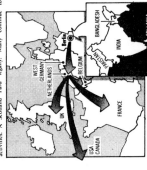

Dutch police were last night investigating Shezad's activities. A Scotland Yard Dutch refugee association say many more live there illegally. Many continue to be obliged to enter Europe

The route the illegal immigrants take to enter Europe

Figure 24 *These examples from The Sunday Times show the sort of graphics that can be used to extend the explanations given in the text*

not scattered about the paper as makeweights in the absence of photographs.

The examples quoted are what might be termed working graphics. Symbol graphics, the drawn or bleached-out motifs used to label a column or flag a running news story such as an election or a big strike have a longer history. They abound on the logos of well-known columnists, holiday supplements and racing cards – a silhouette here, a pair of sunglasses there, the scales of justice, a winning post, pit-head gear and so on.

Symbol graphics are not used to extend the text but to give easy recognition to regular features and looked-for parts of the paper. They are part of the paper's visual character and are, in effect, trade marks, of which the most important is sometimes the paper's own mast-head motif.

The provision of newspaper and magazine graphics is a growing area in computerized printing systems with new symbols being devised to flag stories and, for example, provide story breaks in place of cross-heads.

Alongside this newly developing field, the old-established system of syndicated thumb-nail cartoons and comic strips from independent artists, some with world distribution such as Peanuts and Andy Capp, continues to supply a need for drawn material in a variety of newspapers. Many national newspapers still run a daily or weekly big cartoon on a topical or political subject.

5 SUBEDITING: BASIC TECHNIQUES

The basic editing of copy on a newspaper is carried out by subeditors. These vary in number from three or four on a small weekly, ten to fifteen on a town evening paper, to perhaps twenty or as many as thirty on a big national paper. Because of shift systems and holidays not all work at once. Some papers have separate features subeditors, some divide the news subeditors into a home and a foreign desk. The subediting function, however, is the same.

The job of the news subeditor is to:

1. Check the names, addresses, figures and other facts of the story.
2. Check and put right any errors of grammar or spelling.
3. Check for legality, if necessary referring queries to the office lawyer.
4. Edit the story to the length needed for the page, if necessary cutting the copy or collating copy from a number of sources into one story.
5. Rewrite any part, or all, of the story, as needed, to reduce it to length or bring it to the standard required.
6. Prepare the story for the press by nominating instructions for typesetting.
7. Write a headline, with type instructions.
8. Write any captions needed to go with the story.
9. Revise the story in length and context for later editions in the light of new information or a change in position in the paper.
10. Ensure that all these procedures are carried out to meet page and edition deadlines.

Features subediting is described in Chapters 14 and 15.

GETTING IT RIGHT

The amount of news and the sheer accumulation of facts in a newspaper, coupled with the speed of production, makes inaccuracy a daily hazard.

People are quick to condemn when a newspaper gets things wrong, and the graver errors can get an editor into serious trouble.

A vital part of the subeditor's job is to *check, check, check* everything that is checkable. If avoidable mistakes get into the paper it is not the writer who is blamed but the subeditor who handled the story, who is expected to have almost a sixth sense that detects hidden traps.

Error can arise in many ways. No two eye-witnesses see things in the same light so that discrepancies can occur in different reports of the same incident. Reporters have to be relied upon to cross-check information on the job. There is not much the subeditor can do about this aspect except check back with the reporter any fact or figure that is suspect.

Checking with the reporter should be a regular practice where there is the slightest doubt about anything. Even if the original source of information cannot be contacted by the reporter, a quick word might reveal an error in telephoning or in keyboarding (either typewriter or VDU). A nought might have been added to a figure, or left out, or a sentence garbled. An unlikely age, an unusual spelling of a name, or a figure or sum of money that seems disproportionate to the facts should alert a subeditor to possible transmission errors.

'Attractive blonde model Jenny Jones, aged 43 . . .' would write in angrily if she were only 23.

A man fined £50 for a minor theft would regard it as damaging if the newspaper added a nought to the fine by mistake.

Misreadings can lead to mistakes when copy is re-written by the subeditor. All rewrites should be checked against the original, whether typewritten or electronically keyboarded, especially names and addresses, ages, dates, figures and quotations. It is annoying if the reporter gets the facts right and they turn up wrong in the paper.

Quoted speech is a particular source of complaint to the Press Council, where meanings have been changed or frustrated through vital words or sentences being left out in the editing. The reporter's note or tape is no protection if the quotation has been altered by the subeditor so that the speaker is held up to ridicule or appears to have said something he did not say. This does not mean that quoted speech cannot be edited or shortened, only that it should be done with care, especially in sensitive circumstances.

A common danger is the paraphrasing of quotes to use in the introduction to a story where the subeditor, in his or her zeal to grab the readers' attention, pushes the meaning of a quotation beyond its limit.

If a councillor tells a reporter, 'It is possible the council will consider demolishing three blocks of substandard flats', and the intro says, 'Bromley Council is to consider demolishing three blocks of substandard flats', the subeditor has gone too far. Sources of information can dry up if they get this sort of treatment.

In the offices of competitive national papers, a high degree of polishing, re-writing and striving for effect goes on on the subs' table in the battle to outdo rival papers. The subeditor who combs his or her brains for the telling phrase, the eye-catching intro and words of the right length for a difficult headline count can be in as great a danger of doing damage to the facts as the reporter rushing to get the story in by the deadline.

Cases of *mistaken identity* are another cause for complaint and apology. In captioning stock pictures to go with a story the subeditor must make certain that the person in the picture is identified beyond all doubt. There have been some well known cases in British newspapers where, for example, a perfectly honest, law-abiding citizen has opened his paper at breakfast time to find himself staring out of a gallery of hoodlums at the top of the page, just because he happened to share the same name as someone. Mistakes like this can be expensive.

Picture files often contain more than one person of the same name, and two persons' pictures can become muddled in a file because of this. It is no defence for the subeditor to say the pictures came from the file. If there is any doubt about it being the right person the subeditor should not risk attaching a caption to it identifying it.

Likewise, with pictures of groups, it is not enough for the cameraman to say that he thinks the fourth person from the left is so-and-so. It is down to the subeditor handling the story to find out for certain by whatever means. When the page is made up, the subeditor must ensure that the right captions get on the right picture. The *News of the World* once had the President of the Board of Trade playing outside right for Chelsea.

Picture captions and mistaken identities are two of the biggest traps a subeditor has to deal with. There is no calculating what damage might be being caused to someone's reputation or business by a simple mistake. Even where, through no fault of the paper, a person is mistaken for someone else through having the same name and similar address or description as the person in the page, editors will publish a short *disclaimer*, if asked, if they think damage might be caused.

Mistakes which cause annoyance and irritation to a person or organization, rather than actual damage, can be put right by agreement with the editor by a simple *correction* in a later edition. More serious mistakes can lead to an action for *damages* or, more frequently, the paying of an agreed sum out of court and publication of an apology. Needless to say, editors do not like having their mistakes aired in public and look to their subeditors to protect them from trouble.

In addition to the points made above, there are a number of routines which should be followed by subeditors to ensure accuracy. It is best to check stories against the files in the *cuttings library*, even though the reporter might be presumed to have done this. This will show what previous publicity the person or subject has had and will supply useful background detail which might make the story more intelligible. It will also tell the subeditor if the story is, in fact, new, or has already appeared in another paper. Cuttings can contain names of people who might be used as check sources, and can give biographical facts to flesh out sketchy details supplied by the reporter, or to correct actual errors.

It is always worth checking the *names*, *titles* and other references to people in the public eye. Peers, judges and lieutenant-generals can be touchy about mistakes, particularly the next time you want them to talk to you. And beware about their marital state. Be up-to-date (through cuttings) on divorces and remarriages, and whether people really are married or are only thought to be married. The real wife can be annoyed to find a mistress masquerading in her place in print. Have some idea about an actor's recent performances and a writer's recent books.

Fortunately, in this area checking is not difficult. Titles, heirs, family arrangements and even hobbies of the famous can be found in *Who's Who*; while *Debrett* and *Burke's Peerage* can keep subeditors up-to-date with peers, knights and baronets. It is part of the nature of the titles minefield that Lords Astor, Douglas, Russell, Cohen, Lloyd and Balfour (to name but a few) seem to come in pairs and even threes, and the right one, of wherever he's from, must be identified in the story.

Sorting out Revs, Right Revs, Vens, priests, pastors, rectors, abbots, archdeacons, canons and rabbis, becomes possible in individual cases by reference to such works as *Crockford's Clerical Dictionary*, the *Catholic Directory*, *Free Church Directory*, *Baptist Handbook*, *Jewish Year Book* and others. The word priest is particularly misused in stories about religions.

Whether a school is a public school or not can be gleaned from the *Public and Preparatory Schools' Year Book*, and the names of MPs, their titles, constituencies and qualifications from *Dod's Parliamentary Companion* and *Vacher's Parliamentary Companion*.

For military people and military matters there are:

The Air Force List
The Army List
Jane's All the World's Aircraft
Jane's Fighting Ships
The Navy List

For the Arts there are:

BBC Year Book
Oxford Companion to Literature
Oxford Companion to Music
Radio & TV Who's Who

Who's Who in Music
Who's Who in the Theatre

In business, Government and other areas at home and abroad there are:

Civil Service List
The Directory of Directors
Foreign Office List
International Who's Who
The Stock Exchange Year Book
Who's Who in America

Quotations from the literary greats can give offence when rendered wrongly, especially in those papers that believe their readership to contain a stratum of higher culture. Is it 'all that glitters is not gold?' or 'all that glisters is not gold'? Which homely English poet launched the phrase about tea, 'the cup that cheers but not inebriates'? Is 'Get thee to a nunnery' from *Hamlet* or *King John*? And was it Sir Toby Belch who said: 'Dost thou think because thou art virtuous , there shall be no more cakes and ale?'?

Memory plays false those who think they remember best and the safest thing is to resort to the *Oxford Dictionary of Quotations* or to its Everyman's or Penguin equivalent.

Despite all these prestigious and accessible volumes there are four reference works that are inseparable from any subs' table: *Whitaker's Almanac*, which has the most comprehensive information about governments (home and abroad), trade unions, economies and geographies; *Bartholomew's Gazetteer of Great Britain*, which can settle every dispute about the spelling and location of towns and villages; a modern dictionary and a full set of telephone directories.

Add to these a good international gazetteer and a selection of national year books and you are factually on the mend.

ORDER AND SHAPE

We have seen from the way news is selected, and from the variables at work in news creation, that a great deal more news copy enters a newspaper than can be used. Even after copy-tasting, much remains to be done to reduce the amount of copy to the size and shape needed to projection in the pages. The subeditor is the catalyst for this process. The main task of subbing, after checking the facts, is the cutting and collating of copy so that a story is given an orderly shape within the space chosen for it.

There can be hard feelings between reporters and subeditors over this. Here the subeditor has the advantage of having invariably worked as a reporter before joining the subs' table, whereas the reporter is more concerned with the two hard days spent on gathering a story than with questions of balance on the page.

It is not often possible to give reporters a precise length to which to write because the facts governing space – the number of pages and the volume and quality of news stories available – become known too late. Some favoured news stories might, in the end, get less space than expected, or be dropped altogether. In other cases stories can grow. Extra copy might come in from other sources or a story might become topical because of unforeseen circumstances. Thus the length and shape of a story can be decided hours after the reporter has finished writing it. It has entered the editorial production system and it is the subeditor who is now in control of it.

Having been briefed by the chief subeditor on the length and setting required, with guidance perhaps on the handling of aspects, the subeditor 'shapes' the story to the page. A quick reading of the copy will reveal by how much it needs cutting, if at all.

Some stories might be 'schemed' into the page at almost their original length. Some are the work of several hands, with important aspects embedded in accompanying agency tapes, or with the promise of 'more to follow' from a reporter still covering the job. Some 'running stories' continue right through a day's editions with 'add copy' falling at intervals and revisions necessary.

Throughout this job of checking, cutting and revising, not only of one story but of several during the shift, the subeditor has to carefully assess and update his or her material for the reader, keeping in it what is essential to the reader's understanding of it, but being prepared to modify it in the light of the latest information.

This assessment is based on a close reading of the copy and a judgement of what the story is 'saying'. Newspapers are accused of formularizing news into stereotyped situations. Some criminal court cases, marriage-and-divorce situations and showbiz stories seem to come type-cast, with the order of events and even the things people say being recognizably similar. Such stock situation stories are not common enough to make life easy for the subeditor. Sometimes the crux of a story has to be pulled out of a long account, or comes to light through a query raised. The simplicity and directness of the final version can conceal a lot of work.

There has to be order and shape in the editing of a news story to get the maximum benefit from the space it occupies. The essence of what the story is about has to be drawn out and made into an attractively worded introduction, or *intro*, so that the reader's attention is aroused. Thereafter the facts of the story must be unfolded so that the intro is justified and the story fleshed out. The story has failed if the supporting facts fail to justify the intro or if they leave the reader asking questions.

The facts themselves have to have an order of importance, those most connected with the intro statement being given first, irrelevant ones removed and the least important left to the end. The notion that a story should have a potent end paragraph is an attractive one but in the heat of late copy and quick edition changes a story often has to be cut at the last minute and it is easiest cut from the end. It is also true that the least likely read paragraph in a news story is the last one, so that a pearl of information could be wasted.

Various attempts have been made to impose a neat formula on copy subbing to help in the teaching stage, but none is watertight. There are always stories that require a different approach, and some of these differences of approach will be examined in later chapters. Perhaps the most useful way to look at the technique is to say that a well subbed story should have:

1   *A statement*
2   *Explanation*
3   *Corroboration*
4   *Qualification*

These categories, as will be seen, apply only in the most general way, as do most teaching slogans. It is perhaps easier just to say that the story should have a logical sequence in which the facts drawn to the reader's attention in the intro are explained in the body of the text for the reader so that story is as clear, complete and up-to-date as the subeditor can make it in the space and time available.

Let us now examine the various stages.

THE INTRO

The opening paragraph of a news story must hold the reader's eye once he or she has started to read it so that he or she is persuaded to read on. The intro, as it is called, is a contrived device in which attention is gained not by starting at the beginning of the story, as the first chapter of a novel usually does, but by giving first the highlight, or the most important or interesting part of the story. The subeditor establishes this in the reader's mind and then arranges the facts in sequence to follow.

The fact sequence explains and justifies the intro and amplifies the story so that the reader will read right through to the end to get the full picture. Fashioning a good intro is a skill a subeditor must learn since it is the cornerstone of the method by which news stories are edited and presented in modern newspaper practice.

Experienced reporters, who are aware of the importance of intros, can produce copy that requires little editing, but there are always stories in which the vital fact is not immediately apparent. Later information or the arrival of copy from other sources, or the more experienced judgement of a night editor or chief subeditor, can result in the story being treated in a different way. Its position on the page or the arrival of news pictures to go with it can cause a shift of stress to be given to parts of the story.

With a straightforward story the subeditor might know what the intro should say after a first reading of copy. With a running story or with several copy sources, or with changing news angles, some preliminary tasting and editing is usually necessary. The more complex the story and the more various the copy sources the more variable is the approach to the intro. With a big running story of a disaster it is sometimes the last paragraph to be written and put into type since it must contain the most up-to-date information. In such cases stories are keyboarded into the system for typesetting in sections as edited and are marked 'intro to follow' (see Chapter 11).

Useful guidelines which can help a young subeditor master the art of intro writing are best demonstrated by examining actual examples.

Let us see how a provincial British evening paper, the *Liverpool Echo*, writes its intros. Here is a little story about a missing mother:

> A mum missing for more than two months has contacted her family after the Echo revealed the plight of her three young children.

There is a simple sentence. To be strictly grammatical it should have said 'after the Echo *had* revealed' instead of the simple past tense, but it is a small fault. The point of the story is clear and uncluttered and the reader is intrigued to know about the circumstances.

We read in the next paragraph that teenage schoolgirl Christina Mullen had written to the paper asking for help in tracking down her mother, who had walked out of their home. It gives the address and the date.

Good. To have tried to include these facts in the first paragraph would have caused clutter and made it too complicated.

The dénouement, in fact, is kept to the third paragraph. Christina and her two brothers, John and Steven, have had a letter from their mother to say she is safe.

We would like to know more. Is she coming home? The story does not say. Probably no one can tell at this stage, otherwise the story would have said. It has said neatly what there is to say. The story demonstrates how the simple fact of a letter received, and the circumstances surrounding it, makes news. The subeditor has also highlighted in the intro the newspaper's part in the story. This is good for the paper as well as for the reader.

The story demonstrates two rules about an intro.

- *It should highlight the salient or most interesting part of the story.*
- *It should not try to say too many things.*

Here is another intro which is less effective:

A total of 325,000 children will have their education disrupted next week when the National Union of Teachers stages the fourth in its weekly three-day selective strikes in support of the teachers' pay claim.

It is not a bad intro because it gets to the heart of what is to happen, but it uses too many words. It would have been better to have said '. . . stages its fourth three-day selective strike over pay'. That would have used nine words instead of sixteen for that part of the sentence, and would have avoided using the word 'teacher' twice. The rest could be explained in the second and third paragraphs.

This brings us to another rule about an intro.

- *Every word should count in making it attractive and readable.*

The third example from the same issue shows how, in a town evening paper the 'where' of a story can be an important intro point.

Liverpool Walton MP, Eric Heffer, vowed today to continue his battle for a Labour Party pledge to pull Britain out of Nato.

There follow eight paragraphs of explanation about the issue, and why the local MP feels as he does. This is an important political subject on which parties feel strongly and it is given point by being presented to readers in terms of local connections.

It will be seen in all these intros that two sorts of judgement are called for from the subeditor: one, a judgement of news value in terms of the paper's readership; and two, a language judgement in terms of words and sentence structure by which to present the facts clearly to the reader. If the first story had begun:

Following a letter to the Echo asking for help in tracking down her mother, teenage Christina Mullen has now had a letter from her

it would have read flatly. Why? Because it gives fourteen words of background before getting down to the point of the story, thus giving precedence to the wrong facts. The news judgement is at fault. Worse still, it starts with a subsidiary clause, thus robbing the sentence of pace and verbal immediacy. The language judgement is also at fault.

An intro that starts with a subsidiary clause, or with the words 'following', 'as a result of' or 'because', is putting the cart before the horse. The causes of an event, or what went on before it, though important, are less important than the event that is being disclosed. They should become part of the explanation that follows. This brings us to another important rule about the intro.

- *It should not start with a subsidiary clause because this weakens contact with the reader.*

In making a correct news judgement the subeditor is recognizing the point of the story, which is something that the reporter (who has put together all the facts) may not have recognized. For a story to be chosen to be used it has to have a point to it. It is news in terms of the paper's readers because it is out of the ordinary or of special concern to them, rather than being a slice of ordinary humdrum life. While news has to be the first tidings or disclosure about an event, it has to have something about it that makes it of interest to the reader and worth printing in a newspaper – which is where the whole business of selection comes in.

Even a report of a school sports egg-and-spoon race has its point – the disclosure of who won. The sports themselves have their point – which school house won, and who were the champions among those who took part. It would be nonsense, in news value terms, if the paper carried a long description of the sports and gave these important pieces of information right at the end. The subeditor's news sense is the guarantee that a story will be properly presented to the readers, even if twenty folios of copy from two reporters, a local correspondent and a news agency filed chronologically, have to be searched to find the facts that matter.

Here are two intros that a local paper short of staff and time might have let through:

Local USDAW delegate Len Nobes told the TUC conference that TV publicity about the NGA–SOGAT dispute was damaging to trade unionism.

'I believe the Anglican and Roman Catholic churches will draw closer together as a result of world reaction to the Pope's work for peace,' said the Bishop of Durham, Dr So-and-so, when Convocation resumed at York yesterday.

In the first example the reader knows what TUC and TV stand for, but what about USDAW, NGA and SOGAT? This intro would fail to keep the attention of any reader who is not in the know. It also looks ugly with five capital letter abbreviations in one sentence, even though two can be pronounced as words, though meaningless. This brings us to another rule about the intro:

- *It should avoid the use of abbreviations where possible and either spell out organizations in full or use a general reference, such as the miners or the engineers.*

In mentioning organizations generally, the subeditor should give all but the best known in full at first mention and then, if need be, use the abbreviated form afterwards.

In the second example quoted the effect of the intro is weakened because the reader has to digest twenty-nine words of quotation before finding out who is saying it. While the things people say make news it is usually in the context of who said it and in what circumstances. The quotation used here is meaningless until we find out that it is by a leading Anglican divine on an important church occasion, and scanning through twenty-nine words is liable to make us lose interest before we get there.

A better way would be to rewrite it:

> The Bishop of Durham, Dr so-and-so, said yesterday that he believed the Anglican and Roman Catholic Churches were drawing closer together as a result of the work of Pope John Paul II.
>
> Dr so-and-so referred at Convocation at York, yesterday, to 'the world reaction to the Pope's work for peace . . .' etc.

The intro rule to remember here:

- *It should never start with a quotation.*

Look at the facts in the following story:

> Yesterday afternoon in Salford, a five-year-old boy, Kevin Jones, was knocked down by a reversing lorry outside his home and died later in hospital.

Here is an example of the right information in the wrong order, with the result that the effect is weakened. The intro should relegate the time and place to the end of the sentence and get straight down to the news fact:

> Five-year-old Kevin Jones died in hospital after being knocked down by a reversing lorry outside his home in Salford yesterday.

The intro rule:

- *It should not begin with the time and place.*

The delayed drop

One type of intro that does not follow the above rules is the delayed drop. Here the substance of the story is deliberately kept from the reader to create a feeling of suspense. It can work well in off-beat humorous or human interest stories or in atmospheric stories in which the facts depend more on how they are presented than upon their own importance. For example:

YESTERDAY was Vera Jones's 46th birthday. It was a day she is not going to forget easily for five good reasons.

FIRST, she slipped getting out of bed and emptied her Teasmaid pot of tea all over the carpet.

SECOND, the cat, unused to such confusion, took flight and jumped out of the first floor window.

THIRD, when Vera struggled down to the kitchen she found it was flooded. She'd left the tap on.

FOURTH, when she managed to get to the local pub just on opening time for a birthday drink it was on fire.

FIFTH, when she returned home for lunch, fed up to the back teeth, she found a letter on the mat saying . . . she'd won £10,000 on the pools.

Vera, of Culrose Street, Bolton, said last night . . . etc.

The story breaks all the rules. It starts with the time of the happening. It leaves the main point to paragraph five, and where Vera lives to paragraph six. It uses the shortened forms *she's* and *she'd*. Yet, if told straight it would have lost some of its bite.

Delayed drop intros are effective on the right sort of story but like many special devices they become tiresome with over-use. The story structure must be contrived with great skill so that the reader is 'hooked', and the dénouement must make the slow-burning effect to have been worth while.

On no account should this treatment be used on a hard news story, nor should more than one delayed drop intro appear on the same page. It has to be kept for a special sort of story (see also examples in Chapter 13).

STORY SEQUENCE

Writing an effective intro, when needed, on a reporter's story, or improving the existing intro, not only helps to 'sell' the story to the reader but sharpens in the subeditor's mind what the story is about. This makes easier the job of presenting the facts in logical sequence so that the story has shape and pace as the reader moves through it.

The first point to remember in choosing and ordering the facts is that the intro must not be left unsupported. The paragraphs immediately following it must explain any incomplete references it contains even if this means changing the order of the reporter's copy. If it describes a boy of seven then his name and details, if known, must be quickly given. If it contains a general reference to a union, group, or other organization, then the full identification must be given soon so that the reader is not left guessing.

Any device or generalization used to make an intro short and pithy – references to leader, boss, court, team, minister, etc. – rely upon a fuller description being introduced in the next two or three paragraphs. Where, for example, a story says:

> A Labour councillor told a delegation of young mothers at the town hall today that the party supported their demand for more crèches. . . .

the councillor and the composition of the delegation should be quickly identified.

Thereafter the facts and quotes should unfold in logical sequence. Facts given in the intro should not be repeated as new facts but should be referred to and amplified as they occur in natural sequence. If the story has several 'ends' (information from different locations, or writers, or enlargements of specific aspects of the story), the segments should be connected by linking phrases so that the reader knows that the narrative has diverted from the main story, even though the material still relates to the theme in the intro.

The skilful unfolding of a news story should have a natural rhythm of its own whether it consists of three paragraphs or twenty-seven paragraphs.

Here are some special points about story sequence:

Quoted speech

In stories based upon eye-witness interviews there must be first a proper account of the event, even though an eye-witness's words might be used in the intro. This is necessary to back up the facts in the intro and to give context to the descriptions that follow. Selected interview quotes should then be given where relevant. It is useful to have a number of accounts to corroborate each other. For instance a story of a train struck by lightning might begin:

> A hundred people were injured when an express train ran off the lines north of Watford after being struck by what a witness called 'a ball of fire'.

Later, after the main facts, the story might say:

> Miss Jane Watkins of North Road, Rickmansworth, who was in the third coach, said: 'There was a great flash as we passed over a bridge and everyone screamed'. Her friend June Smith of Long Lane, Welwyn, Herts, said: 'The fireball seemed to land right on the front of the train'.

- A story based upon what someone has said is incomplete, and can lead to trouble, if the person's quoted views are not included as they were said. If a story starts:

> Boomtown's Tory mayor said last night that the Labour Opposition group on the council were 'preying' on people's misfortunes to make political capital. . . .

then it should have a qualifying paragraph to justify this:

> The Mayor, Coun. Thomas Jones, declared: 'The Opposition on this council are like vultures . . . etc.'.

- In introducing quoted statements by people, beware of misrepresenting their tone by using, or leaving in, words such as 'criticized', 'insisted' or 'demanded' in an attempt to sensationalize them or create headline ideas. Such verbs require strong views to sustain them. If a councillor said, 'I think house-building tenders should be advertised over a wider area to try to bring in cheaper quotations,' he could be aggrieved to find on reading the newspaper report that he had 'denounced' his council's house-building programme.

Many adjectives and verbs used in describing the things people say can have a loaded or emotive effect which goes beyond the objectivity required in news subediting. There is nothing wrong with the neutral 'said' or 'told' or 'added'. Quotations have to be selected to suit the space and context, but at least let the person's words speak for themselves.

- Some subs, in an attempt to get everything of interest near the top of the story, break up interview quotes used by the reporter, using first those of one person, then those of another, and even introducing a third speaker in the first few paragraphs. This might be useful for a factual order the sub has in mind but it can confuse the reader. In cases of reported discussion, the mixing of quotes and speakers is justified, but in the run of a story it is best to deal one at a time with the things people say rather than break up and shuffle their words. There is also a danger that if parts of an interview are detached and used in the same paragraph as other people's words their original meaning and intention might become distorted.

Where statements are made in a story, or by someone interviewed in a story, involving or criticizing another person, then the other person should be given an opportunity of commenting on them. If the reporter has not done this then the subeditor should go back and ask that it be done. Even if the person says, 'I have no comment to make' it is worth using it to show that the newspaper is trying to be fair.

- Whether quoted or paraphrased, statements in a story *must* be properly attributed.

Here is a report of a political speech from the *Guardian* morning paper, based not upon interview quotes but upon quotes from the actual speech, which demonstrate some of the above points:

By James Naughtie, Chief Political Correspondent

The inner city crisis was blamed on spiritual and moral problems rather than economic difficulties, by Mr Norman Tebbit yesterday, in another rallying call to the Conservative right-wing.

The Tory Party Chairman said it was the Government's duty to restore order – partly through tougher sentences as a deterrent to crime – and to overcome 'the poisoned legacy of the permissive society'.

He said that even if the physical structure of neglected urban areas could be transformed, it would be in vain unless personal responsibility was restored. This came through ownership, security, established by law and order and family stability, he said.

In an address at St James's Church, Piccadilly, London, Mr Tebbit said that the freedom he sought for the country had to be buttressed by the law and by social pressures, notably a concern for standards and a sense of personal responsibility.

Identifying the root of the nation's problems as moral, he said: 'British society must regain a sense of order. Order reinforced by punishment of violent criminals. Order in our streets. Order in schools and order in the home'.

He went on to accuse those who regarded public feeling on law and order 'with a certain lofty disdain'. He said that the violence in inner cities had been engendered not only by physical but by moral decay – a lack of self-respect stemming from the collapse of traditional values.

'In my view too many well-meaning people concentrate solely upon treatment of the symptoms of inner city decline and fail to see that miseries derive in great part from family disintegration and its causes – which are often spiritual and moral, not material,' Mr Tebbit said.

He argued that freedom – both economic and individual – depended on values which had been eroded by the permissive society, which he said had progressed from the 'anything goes society' to the violent society.

'Legislation on capital punishment, homosexuality, abortion, censorship and divorce – some of it good, some of it bad, but all of it applauded as progressive – ushered in quick succession an overwhelming impression that there were not only going to be no legal constraints, but that there was no need for restraint at all,' he said.

Mr Tebbit argued that the nation was rightly concerned about inner cities, but the failure of previous collectivist policies was shown in the deprivation, rootlessness and poverty which went alongside a lack of self-respect.

The intro summarizes the thrust of the Tory Party Chairman's message allotting the blame for the 'inner city crisis', linking it in paragraph two with the party's policy on law and order. There are four paragraphs of paraphrased material giving the readers the gist of the speech, followed by a telling quote in support in paragraph five. A few introductory words in paragraph six take the reader into further quotes about the public's attitude to law and order, followed by closing paragraphs of paraphrase. The occasion of the speech is considered of less consequence than its content and is brought in at paragraph four.

The juxtaposing of paraphrase and quoted speech presents a rapid, unfolding pattern of the substance of the speech, where quotes alone could have slowed it down. Political speeches do not normally present their most potent points in the first few minutes. Rhetoric is usually allowed a place in the leading-up process. The treatment here is a good example of how a newspaper, by identifying the news point and encapsulating it in the intro, and thereafter arranging the sequence in order of fact importance, can take often wayward political word patterns and turn them into news.

The second example demonstrating intro and sequence is from the *London Evening Standard* and is about local electioneering. Here again the intro isolates a telling point in a party's campaign which it feels will be relevant to readers who live in the area:

SOCIAL Democrats fighting to win control of Islington Council from Labour have said they will turn the borough into a 'squatting-free zone'.

They say that anti-social occupants who make their neighbours' lives a misery will be evicted if the SDP wins the elections on May 8.

David Hyams, leader of the council's Social Democratic Opposition, said: 'Labour has turned Islington into a squatters' paradise. Their very first action after winning power in 1982 was to declare an amnesty for squatters'.

Their manifesto says: 'As council housing becomes vacant it will be immediately allocated to new tenants. Any squatters or queue-jumpers in these properties will be pursued actively through the courts until evicted.'

Labour councillor, Chris Calman, chairman of the housing committee, denies the council is a 'soft touch' for squatters.

He said: 'When we find out that someone is living in a house without permission, we go to the courts for a possession order. We also make use of the criminal law in appropriate cases'.

Paragraph one sets up the story. Paragraph two qualifies and explains the point about the party's intentions of dealing with a borough's squatters. Quotations in paragraphs four and five are the justification for the story, and the remaining two paragraphs give the Council's spokesman the right to comment, demonstrating that the story is not just a piece of party propaganda.

Geography

The local connection of a story is important in a paper serving a local or regional area and is a valid intro ingredient. In a national paper, geography is less important and is often relegated to the third or fourth paragraph or even later. Readers can become annoyed, however, if the location of a story is kept from them until almost the end when they find to their disgust that it has happened not in Little Tolpuddle down the road, but in Ulan Bator, Outer Mongolia.

Even in national papers, geography in a national sense matters, and foreign stories are often used in terms of their British connection. A person involved in a story about Saudi-Arabia would be a Briton in paragraph one, while he would be 'from Halifax' or wherever, in paragraph six or seven. In Halifax his local connection would merit the first paragraph.

Time

The *yesterday* or *today* should be made clear in the intro and followed through in the story. Any changes in the time sequence should be introduced by a linking phrase (sometimes in italic) such as, 'earlier in the day' or 'speaking last night'. Once the time sequence has changed, the story should not revert to the original time or the reader will be confused. Other 'ends' to the story should be introduced after the main explanatory text by a similar device such as 'a spokesman at the town hall said later,' or 'Mr X's former wife told reporters by telephone from New York'.

Background

In giving background material, especially from press cuttings, records and reference books, subs should beware not to use it to slant a story or influence the reader's reaction.

A person's career or achievements can often be filled out usefully in the editing, but if a person appointed to public office has been found in cuttings to have been fined for theft when a teenager this could be actionable if used, even though true. It could be construed as malicious (see Chapter 10).

Any background material worked in must be relevant to the story and not be used to support a point of view.

CASTING OFF

The usage 'to cast off' comes from a hot metal printing term meaning to set a piece of copy or a headline exactly to fit a line of type. In subbing, it is widely used to mean to edit copy to an exact length to fit the space on the page.

To cast off a story, the subeditor must first know how many words can be contained in a given column space in the newspaper. The best way to work this out is to cut specimens of body type, 7 pt, 8 pt and 10 pt say, from the paper and stick them on a card in 1 inch or 3 centimentre amounts. Metric measurements are now more commonly used in electronic editing. The examples of Excelsior type illustrated (Figure 25) are taken from a tabloid-size page of standard 8½-em columns. Word count will vary according to the body type used and the different column widths. A check card should show column widths, double-column and column-and-a-half.

The number of words to the column centimetre (or inch) in any given body type quickly becomes familiar to a subeditor, although a count might have to be made in the case of non-standard (i.e. bastard) setting where the depth to be filled is critical.

It is equally necessary to know how many words a given story contains when it is submitted for subbing so that the right amount of cutting can be made if it is too long for the space. On screen this is no problem since the header will give the word count of the copy when entered. Where hard copy subbing is still the practice, or with any off-screen editing, the length can be quickly checked by counting specimen lines on a page of typescript, or print-out, averaging them out and then multiplying them by the number of lines on the page. An average number per page can be worked out by the same means and this can be multiplied by the number of pages to give the total number of words. For instance, this book has about ten words to the line and an average of forty lines to the page (excluding pages with illustrations or ends of sections). Therefore each page contains approximately 400 words. If turned into newspaper type these words would fill a 20-centimetre depth in 7 pt Roman type in an average 8½-em column.

A practised eye can measure the initial length of texts, whatever the number of pages, without being more than a few words out and should likewise be able to estimate how much of it can be contained in a given space on a page. The big problem of getting an accurate cast-off for the page in hard copy subbing arises where there are heavy editing marks and deletions which disturb the balance of the lines. This is no problem on screen since the text adjusts and rearranges itself in front of the subeditor and a word and line count can be obtained at any time at the press of a command key.

The actual reduction is achieved by first considering the *facts* of the story. Those that are needed for the intro and main sequence are left, while those of least importance, which are either irrelevant to the points that need to be

1-inch   I hated it but there wasn't much I could do about it. Just before we got together again he told me this woman only had a year to live and   10 pt Roman lc × 8½ ems

1-inch   The new passion in Jim's life, who has rescued him from the depths of despair, is his own daughter Sarah, from his first marriage.
    Before their recent reunion they had not   8 pt Roman lc × 8½ ems

1-inch   and when he said he might take up a permanent suite he got the real red carpet treatment."
    In fact, the cunning count was broke—but he even conned the Ritz into letting him take out £500 cash advances saying he could not be bothered to   7 pt Roman lc × 8½ ems

Figure 25 *Column width setting by the inch – a useful paste-up guide for casting-off in text editing*

made or for which there is going to be no space, are deleted at the start. The remaining material is then checked and rearranged where necessary, making sure it reads properly where deletions have been made, to give the story its *shape*.

This is the part of subbing in which the subeditor, reading and deleting, is mentally putting the story together around its salient points, basing it upon a personal news assessment and a knowledge of the newspaper's readership.

The next stage is to work through the copy carefully, paring down the text, rooting out repetition, ambiguity and redundant *words*, shortening wordy phrases, checking the grammar where necessary and gradually tightening up the structure so that every word does a job and the best use made of the space.

Not all copy needs such heavy attention to the words as this, but the subeditor, faced sometimes with a small space and a good deal of copy, cannot afford to waste words or allow slow, cumbersome grammar to get in the way of the story, or eat up the space.

At the end of this stage the story should be somewhere near completion, and a final reading of the material should enable the subeditor to make a few adjustments to fact and word needed to bring the copy to the required length and standard. This part of the job is the second most important after checking for accuracy. Bad casting off, either well under or well over the required length, can cause much trouble because of the amount of late cutting or adding needed to make a story fit, and delays can result in completing pages.

As will be seen from points raised in the next few chapters, the economic use of words is essential in using space to maximum effect to enable modern newspapers to carry the number of news stories they do. Tight subbing should be regarded not just as using fewer facts and details but as making way for more by pruning inessential words and phrases. It is the content of a news story that matters to the reader.

The illustration Figure 26 shows how a subeditor, working on hard copy, tightens up the text to make more space for essential facts.

ELECTRONIC EDITING

With direct input of copy into the computer the subeditor has the advantage of electronic aids in editing. As with copy tasting, the ease of operation is greatest where everything comes through the computer, rather than with some subbing still having to be done on hard copy. Subeditors who have become familiar with screen editing do not usually like to go back to subbing hard copy.

Stories can be checked for length by relating the word count and number of lines given in the screen 'header' to the space allocated on the page layout. Fact and word pruning are then carried out as with hard copy subediting. The difference is that the use of the cursor, or electronic pen, on screen enables deletions and alterations to be made faster than by ballpen, while at the same time resulting in instantly clean copy without the scribbles, scorings and connecting lines that characterize hard copy subbing. After each change or alteration the text automatically rearranges itself on the screen.

Split screen, on which two stories are displayed side by side, can be used when working on copy from another source. Deleted material may be also left in the computer in 'note' form from which it can be reactivated and brought back into the text, if needed, by a simple command stroke. With electronic editing, copy is capable of almost endless reworking and revising – as it is with any word processor – while at the same time finishing up clean and ready for typesetting. Moreover, after a story has been checked and cut to fit it can be scrolled back for re-reading on the screen and any late alterations made with greater ease than would be the case with already heavily subbed hard copy.

The final stage by which, at a stroke, a story is hyphenated and justified in the computer ready for the photosetter, produces an accurate length and word count in the header on the screen. This enables the subeditor to make any cuts still needed to cast it off exactly to fit the space on the page, thus saving time at the make-up stage. After setting, stories can be recalled to the editing screen for attention to text or setting, if needed as a result of page changes or late news, and be then put back through the photosetter in a revised version.

Modern systems have a facility for delivering print-out proofs after the H & J stage showing the story as it will look when typeset. These are useful if the story needs to be taken away – for telephone checking, say. In fact, many newspaper offices make little use of the facility once direct input has transferred the whole editing operation into the computer. Nor is the print-out proof needed to increase the distribution of a story within the editorial department since it is now possible to generate 'carbon' copies of a story within the system. This means that it can not only be read by one person while being edited by another, but can be edited on screen into different versions for use in different papers fed from the system.

A difficulty in writing about electronic editing is that the sequences cannot be illustrated visually for the reader. The problem is not a serious one since, once the keyboard techniques have been mastered, the sequences are the same as in hard copy subediting. Marks on text are simply replaced by keyboard commands, with the text rearranging itself in word sequence and length on the screen in response to the editing procedures. On screen the instructions follow the same sequence as with hard copy subbing.

Computerized systems do not call for a change in techniques but, rather, bring electronic facilities to the aid of the subeditor in carrying out the techniques, with the aim being to produce the same result on the finished page.

MARKING UP

A basic part of the subeditor's job is to mark up copy for typesetting. In many ways this part is more important in electronic editing than with hard copy editing since there is no intermediary to correct a wrong setting instruction – even if it is a nonsense – once it leaves the subeditor and enters the photosetter. All copy, whatever the editing process, should be marked at the top with the *page* and *edition* and the *subeditor's name*. It must also be given a *catchline* (US: slug-line) so that it can be identified in the system. With hard copy feeding to a keyboard operator, the catchline *gem* 1, *gem* 2, *gem* 3 etc., with each folio being marked *more* and the last folio marked *end*. See Figure 26.

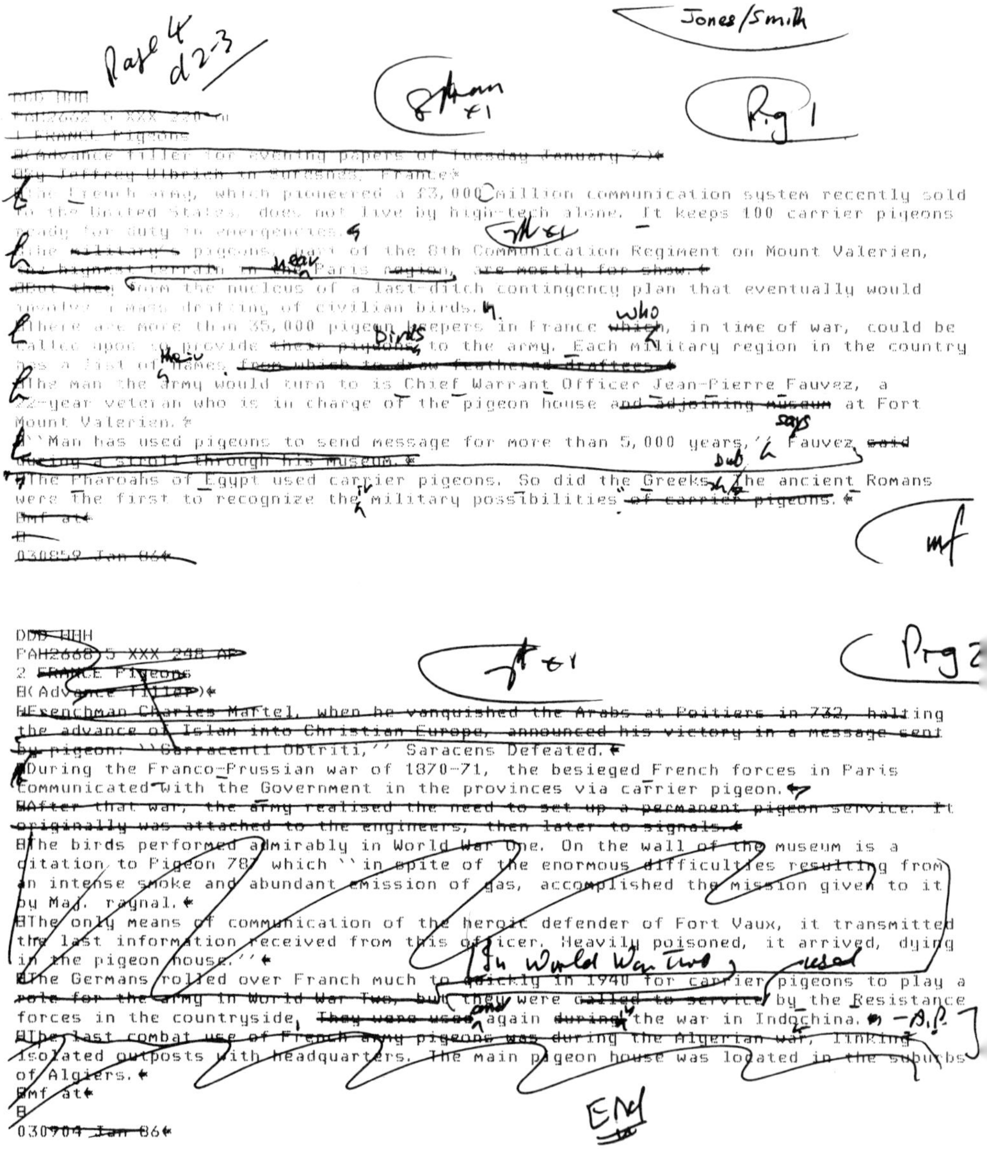

Figure 26 How a story is subedited on hard copy – in this case a computer print-out. With direct input and full utilization of modern systems all these alterations and instructions are keyed in by the subeditor, using a terminal, and the text adjusts to the required wording and length on the screen ready for setting

While folio numbering is not usually necessary with screen-edited copy, the catchline most certainly is, both for the main text and any *adds* or *inserts* that are keyed in. The use of the catchlines is the only way text can be called from the system once it is entered, and each story catchline should be written on the page layout so that the subeditor, either on the desk or in the page make-up area, can recall the story on to a terminal for any adjustment or revision needed after editing. Lost or forgotten catchlines can cause serious delay when stories inside the computer are wanted for revision or re-running through the photosetter.

The *typesize* and width (*measure*) of setting must be correctly instructed on all text before it leaves the subeditor. Intro sizes are usually a size or two bigger than the rest of the text and sometimes of a different measure. Any special setting such as *drop letters, italic* or *bold* paragraphs or *indented* setting should be clearly instructed by the use of the appropriate command keys. Headlines, crossheads and any captions should share the same catchline, or derivative of it, and the same page and edition number as the story to which they relate, so that they can be identified at the page make-up stage as text bromides pour out of the photosetters.

Photosetters are formatted to give certain types and measures on 'programmed' keys, and also by-lines and fixed-style crossheads, to make for speed in keyboarding instructions. The command might be a coded one – not the name of the type but a lettered or numbered code. Any story wrongly or inadequately marked will come to the page in the wrong setting or shape and have to be set again. Accurate marking up is vital to good edition times. The computer will set only what it is instructed to set and serious delays can be caused by errors that are spotted only after typeset bromides have been delivered to the page. The main reasons for such errors are failure to define accurately what setting is needed, or to record properly changes in setting within the story – for example to redefine body setting after inserting crossheads or italic paragraphs.

Marking up text in electronic editing is, in fact, easy and logical provided the routine is accurately followed and the computer is instructed precisely what it has to do. The use of programmable keys for certain setting can remove the need for endless editing marks and quicken the processing of copy. This, added to faster setting and easy recall and revision, should result in better edition times and even later copy deadlines for editions.

6 SUBEDITING: LANGUAGE AND ITS TRAPS

Text editing demands from the subeditor not only the ability to present facts in an orderly way, and with economy of words, but also the facility to spot a variety of errors and misuses that can crop up in the wide range of copy handled. Some of these faults are grammatical – even the best writers go through life with blind spots that were never eradicated when young – and others to do with the use and meanings of words.

Newspapers respond to new usages that are part of a growing world use of English. They reflect popular expressions and 'vogue' words and the subtle shifts of meanings that stem from the spoken language and the changing requirements of commerce and technology.

To pick a way through this minefield and decide when a new usage is acceptable is more difficult for the grammarian than for the journalist, since the grammarian feels himself or herself to be the custodian of the language. The journalist is merely using it to an end.

Even so there has to be some consistency and acceptability in the way language is used if the journalist is to communicate meaningfully with the reader and not be misunderstood. For the journalist the yardstick of success is the effectiveness of language in this act of communication.

This is vital in editing when texts, even though well written, have to be condensed without loss of meaning, words and phrases substituted to save space, ambiguity excised and clarity and order imposed upon a story which might have a variety of copy sources.

The very act of selection and condensation can lead the unwary into error when time is short. In language terms the subeditor is saddled with a dual role of imposing his or her own creativity on a story and acting as quality controller of the whole. Stories cannot be left with difficult sentences, misleading punctuation or words that do not make clear the writer's intentions. The subeditor must ensure that the text on the pages gets through to the reader in an attractive, readable form, and that the best possible use is made of the space allocated to it.

The following guidance on language is given with this purpose in mind.

THE SENTENCE

Good sentence structure is the key. All sentences must have a subject and a verb. The verb, if need be, can be qualified by an adverb:

subject	*verb*	*adverb*
the policeman	walked	quickly

The sentence can also have an object:

The policeman walked quickly towards the boy (*object*).

Newspapers, and some fiction writers, allow the subject to be implied in certain cases, and sometimes even the verb. This can work, provided the sense makes the subject clear to the reader:

The policeman walked quickly. Too quickly (*he walked*).
They did everything to give him a day full of activity. How full a day (*they gave him*)!

The use of an *implied subject* is a device that can give pace to narrative provided it is clear who or what is the subject. It should be used only to produce a special effect. If over-used, or used without justification in place of conventional punctuation, it becomes a tiresome mannerism and ceases to achieve an effect.

Where a verb is inactive (where it expresses a state of being or feeling) it takes a *complement* in place of an object. For example:

subject	*verb*	*complement*
the girl	felt	happy

It is important to know which is the subject of a sentence because the verb must agree with it in person and number. In person the verb changes only in the third person singular:

I *move*   He, she or it *moves*   They *move*

Agreement in number, however, can cause problems. Where a sentence has a double subject the verb must be plural, as in:

Oil and water *do not* mix.

Where words are joined to the subject by a preposition the subject remains singular, as in:

Iron, with copper, *is* the most important metal.

In neither-nor or either-or sentences the verb agrees in number with the subject nearest to it, as in:

Neither John nor his brother *is* a member.
Neither Helen nor her sisters *are* going.

Beware of pitfalls with numbers and quantities. Use *fewer than* for numbers as in:

There *were* fewer than fifty copies left.

Use *less than* for amounts or quantities.

There was less *than* a quorum so the meeting was abandoned.

Generally, numbers used as terms of measurements are singular:

There *was* ten pence in the hat, but: There *were* ten pennies.

None as a subject generally means *not one* and should be treated as singular.

Thirty years ago there *were* many fishing families in the village. Now there *is* none.

Agreement in person and number with the subject is most difficult in the case of *collective nouns*, the term given to collection of things such as a herd, a class, a bevy, a gathering, etc. Grammatically, they should be treated as singular.

The class *was* too lively for the new maths teacher.
The Government *has* decided to scrap *its* proposed wealth tax.

This is a useful rule on the whole, but sticking to it pendantically can lead to stilted structure, and some newspapers allow collective nouns such as the Cabinet or football and cricket teams to be treated as plural subjects to make for ease of reading. In such cases consistency should be the rule. It would be wrong for a subeditor to allow this sentence:

The Government *has* decided to scrap *their* proposed wealth tax.

Words like politics, mumps, graphics and acrobatics should be treated as singular, despite ending in an *s*.

Participles

Participles such as having, going, running, turning, etc., are said to be left dangling in sentences in which the writer has failed to identify the subject. For example:

Having addressed the meeting for two hours, an interval was then agreed.

The subject is the person addressing the meeting, but who was it that agreed the interval?

Passing quickly through the agenda it was then the turn of the treasurer to give his report.

Was it the treasurer that passed quickly through the agenda?

A dangling participle is a certain cause of confusion to the reader and a fault that crops up in the copy of inexperienced writers.

Pronouns

The main fault in the use of pronouns occurs in constructions in which the writer has failed to relate them properly to their antecedents. For example:

Helen cooked for *her* sister *her* favourite meal.

Whose favourite meal?

The general looked at *his* aide grimly.
His eyes were half closed.

Whose eyes?

Where there is any doubt about the identity of possessive pronouns (*his*, *her* or *its*), or personal pronouns (*me*, *you*, *him*), the sentence should be reshaped or split into two sentences.

Confusion of *me* and *I* is a pitfall. Broadly, the rules are these: where the first person pronoun is the subject of the sentence it should be *I*. Where it is the object of the sentence it should be *me*. Thus 'It affects *you and me*' is correct; but you would say '*You and I* are good friends' because *you* and *I* become a double subject – *you* are a good friend and *I* am a good friend.

I should be used and not *me* following a conjunction, as in:

'He is as baffled as *I* (*am*)' not 'as baffled as *me*'
'He is younger than *I* (*am*)' not 'younger than *me*'

Confusion between *who* and *whom* is more deeply ingrained. As a very loose guide, one says *to whom, from whom, by whom, of whom* and *than whom* using *who* in most other cases. Thus, *whom* is used in the sentence when it is preceded by a preposition. A trick is to substitute *he* for *who* and *him* for *whom* in your mind and see how the sentence works out. 'I did not know *he* was the one' would indicate that 'I did not know *whom* was the one' could not possibly be correct. And so on.

A difficulty with pronouns is highlighted by the following passage: 'When the young journalist starts his or her first job it is necessary that he or she should learn to take a competent note. If any problem crops up, especially over court hearings, he or she will be asked to substantiate . . . etc.'.

The glaring fact shown here is that the English language lacks a singular pronoun of common sex that can be used where the references are to either sex.

Otto Jesperson, in *Growth and Structures of the English Language*, refers to the three available makeshift alternatives – either using *he or she*, *they* or just the universal *he* in general references. Sir Ernest Gower, in *The Complete Plain Words*, is not happy with *they* or *them*. 'Each insisted on their own point of view, and hence the marriage came to an end.' He says this usage is not defensible, though he concedes that 'necessity may eventually force it into the category of accepted idiom.'

Strunk and White, in *The Elements of Style*, say boldly: 'The use of *he* as a pronoun for nouns embracing both genders is a simple, practical convention rooted in the beginning of the English language. *He* has lost all suggestions of maleness in these circumstances.' They go on: 'The furore recently raised about *he* would be more impressive if there was a handy substitute for the word. Unfortunately there isn't.'

Quite! Meanwhile this writer and others do all they can to recast sentences so as to ward off the dreaded choice.

The personal pronoun *one* can also be a problem. *One* should follow through, once having started, though *one* can quickly find *oneself* wishing *one* had used a different pronoun, however much *one* feels that *one* should stick to *one's* guns.

Which or *that* – an old bogey in sentence construction. A general guide is that *which* must be used in a commenting clause. Example: 'This should go to the news desk, which deals with these matters.' In a defining clause *which* or *that* is correct as in: 'The committee *that/which* deals with the matter has been disbanded.' (See also under 'punctuation'.)

'The man *that* deals with the matter . . .' and 'The man *who* deals with the matter . . .' are also equally correct in a defining clause.

Sir Ernest Gower favours the use of *that* where the choice is justified as making for a smoother sentence. He also says that either should be dropped in sentences that sound right without, the ear being the guide. This is useful advice. A sentence such as, 'I think that the record which he wants is the one that is in that box,' can thus be trimmed to say, 'I think the record he wants is the one in that box.'

The sentence reminds us of the awkward fact that the word *that*, unlike *which*, can serve as a conjunction, a relative pronoun and an adjective.

Tenses

The necessary thing is to be consistent in the use of tenses. Beware of captions which are in the present tense. Try to avoid:

Elizabeth Taylor *arrives* at London Airport when she *came* to attend the premiere of . . . etc.

If an interview is in the present tense, do not allow *she says* in one part of it and *she said* in another.

Do not mix past and past perfect tenses in one sentence:

'I went there because I have been thinking that I should like to see her.'

A muddle of tenses like this is best resolved by recasting the sentence:

'I went there because I thought I should like to see her.'

The verb *to be* can cause complications in the lesser used tenses. Note that *I was* and *he/she was* in the simple past tense become *(if) I were* and *(if) he/she were* in the past subjunctive. The *were* in this sort of usage should be reserved for unlikely or conjectural situations. For example: 'If I *were* the Prime Minister.' The *if* is not necessary in a sentence such as, 'Suppose he *were* the Prime Minister. . . .'

A good tip is that the tenses of verbs should relate to the tense of the governing (introductory) verb of a news story, which is usually in the past tense. Allow for differences of tense in quoted speech but, at the end of the quoted passage, revert to the governing tense. An exception to this would be any reference to a permanent truth. For example, 'He said that the world *is* round,' is preferable to 'He said that the world *was* round.'

On the whole, try to keep tenses simple. Recast sentences where there are complications. Be consistent.

Verbs

The worrying thing about verbs is the rate at which new ones are being formed from nouns – containerize, hospitalize, civilianize, servicize, computerize and so on. Many of the *ize* (or *ise*) verbs enable complicated things to be said briefly and should not be rejected, but the subeditor should guard against using so many of them that a story begins to read like official, legal or technical jargon.

It is up to the grammarians to decide how many of these new creations are to be accepted into the language, but the subeditor could strike a blow for sanity by, for instances, using moisten for moisturize, complete for finalize, use for utilize and curse for anathematize.

News stories have more pace and immediacy if verbs are used wherever possible in the *active* voice rather than the *passive*. For example: 'The policeman saw the boy. . . .' not: 'The boy was seen by the policeman.'

For the use of the verb in headlines see Chapter 8.

Two constructions seem to give some writers trouble: the use of *shall* or *will* in the future tense of verbs, and the use of *lay* or *lie* in the verb to lie (down).

Shall is normally used in the first person singular and plural as in 'I shall . . .' and 'We shall . . .', while second and third person singular and plural take *will*, as in 'He will . . .' 'You will . . .' and 'They will . . .' This order can be reversed in emphatic statements such as, 'I will go!' or 'You shall win through!'

In the verb to lie (down) say 'He lay down . . .' (intransitive) but 'He laid down rules . . .' (transitive). Say 'You should *lie* down . . .', not 'You should lay down', unless the use is transitive: 'You should lay down rules.'

Split infinitives

The infinitive of a verb is its basic form: to be, to go, to take, etc. Splitting the infinitive means inserting an adverb between 'to' and the verb, as in to quickly go, to quietly take, to always be. People are no longer scandalized by the breaking of the old grammatical rule that infinitives should not be split in this way as they were when H. G. Fowler first sanctioned it in certain cases in his *Modern English Usage*.

Conveying the correct meaning is the important thing, yet this can usually be done without splitting the infinitive. 'To further improve the working of the engine,' makes just as much sense, and makes for a smoother sentence, if worded: 'To improve further the working of the engine.' If we mean there is a further intention to improve the working of the engine, then 'further' must go before the infinitive as in 'further to improve'. To say that the purpose of a drug is to better deal with hay fever, is more correctly rendered as 'to deal better with hay fever', or even 'to deal with hay fever better'.

There is also the question of stress and scansion in some writing. 'For man to boldly move towards his future . . .' has a metrical ring absent from, 'For man to move boldly (or boldly to move) towards his future . . .'. On these grounds – though they are unlikely to affect most newspaper writing – the infinitive might reasonably be split. In the vast majority of sentences the separating adverb can be placed outside the infinitive as shown above, with little or no damage to the sentence, or danger of ambiguity.

Qualifiers

These are words that vary or extend the meaning of the noun or verb and they are of two sorts – *adjectives* and *adverbs*.

Adjectives, which qualify the noun, are frequently the most expendable words in a text since an excess of them slows down the writing and takes up space.

Many, such as luxury flat, stunning blonde, staggering sum, hush-hush inquiry, vital clues, and burly policeman are among the most tired and over-used words to be found in newspapers. Many of the more popular ones, as in surprise swoop, brutal murder and secret hideaway, are plainly as unnecessary as they are tautological.

Another area of popular and over-used adjectives is that based upon nouns, such as miracle, shock, model, terror, horror and love. The more popular national dailies are awash with miracle babies, mums, dads and grandads; shock reports, disclosures, results and endings; model babies, mums, dads and grandads; terror flights, terror drives, love nests, love children and horror stories of many types.

It is in the use of adjectives that newspapers, instead of expanding the frontier of language, seem to be shrinking it. It is a useful exercise for a subeditor, as well as a writer, to cast out all the above adjectives and go for something that precisely qualifies the noun or, if nothing comes to mind, go without. Twenty adjectives saved in a column means an extra one-paragraph news item at the foot of the column. Over the page this could allow in one important extra news story.

Here are some usages that could be pensioned off with little loss: considerable difficulty, serious danger, all-time record, grateful thanks, cherished belief, psychological moment, broad daylight.

One fault to look for is the tendency in some writers to limit or extend absolute adjectives. There are no degrees of comparison in absolute, basic, essential, final, ideal, unique and ultimate. Almost unique or partly ideal are a nonsense. You cannot be more basic than basic.

The use of adjectives/adverbs, such as *very, rather* and *quite* can be debilitating to the text and read like an extension into writing of verbal props. The high incidence of *quite* is (quite) extraordinary.

Beware of *non* prefixes. Non-essential, non-cooperative, non-aligned and non-active invariably mean inessential, uncooperative, unaligned and inactive, while non-professional usually means amateur. The shorter alternatives also have no intrusive hyphen.

In general the rule with adjectives is that they must be doing a job if they are to be allowed to stay.

Where adjectives qualify the noun, adverbs qualify the verb and other adjectives by words such as occasionally, rightly, severely, half-heartedly, simply, implicitly, purely and finally. Be certain to put the adverb in its right position. 'You can quickly teach dogs to do tricks,' is not the same as 'You can teach dogs quickly to do tricks'. Be certain of what you are trying to say.

Nouns

These offer fewer problems than do any other ingredients of a sentence. The use of collective nouns is covered above under 'subject', and of possessives later under 'punctuation'. Plural forms of nouns can give difficulty. The *ey* endings as in valley and money take an *s*; *y* endings without a preceding vowel change to *ies*.

Oes and *os* endings: monosyllabic words like no and go become noes and goes; the commoner two and three-syllable words such as tomato, potato and hero take oes. Long words, particularly imported ones such as archipelago and gigolo, and also proper nouns such as Lothario, all take os endings. Words with a vowel before the *o* such as cameo, intaglio and imbroglio take os, as do abbreviated words such as photo.

Foreign words: many Anglicized ones, among them sanatorium, syllabus, terminus and ultimatum, take a simple *s* or *es* ending. Some French- and Latin-based words, however, keep their own plurals. These include:

addendum	addenda	fungus	fungi
beau	beaux	memorandum	memoranda
bureau	bureaux	minimum	minima
cactus	cacti	phenomenon	phenomena
criterion	criteria	plateau	plateaux

Beware of words such as medium which becomes mediums for clairvoyants, and media for methods of mass communications; series which remains series; fish which can be fish or fishes, and folk which can be folk or folks.

A blight that afflicts current English is the growth of polysyllabic nouns derived from the new generation of 'ize' verbs, themselves often derived from shorter nouns. Thus we get:

container	containerize	containerization
hospital	hospitalize	hospitalization
moisture	moisturize	moisturization

With casual, casualize, casualization, comes a more advanced growth: de-casualization. We are only one step from de-containerization, de-hospitalization and de-moisturization.

The space saved by using such monster words is wasted if the reader is lulled into insensibility before the end of the sentence. It is better in such emergencies to take up a bit more space and use a few short simple words to explain what is meant.

Prepositions

The rule never end a sentence with a proposition is frequently broken without any ill effect in sentences such as 'She's the wife he goes home to' or 'She's dating the chap she works with,' which would sound pedantic as 'She's the wife to whom he goes home,' and 'She's dating the chap with whom she works'.

A good tip with prepositions is that if the sentence sounds right and is free of ambiguity then the preposition can be left at the end. Avoid collections of prepositions at the end such as in the classic, 'This was the book he wanted to be read out of from to'.

Most uses of prepositions with nouns and verbs (conform to, connive at, taste of, taste for, consequent upon, etc.) are idiomatic and have to be learned.

Other points

- Sentences can be made a nonsense through a misplaced clause or phrase: 'For the third time a baby was trapped in a washing machine at. . . .' Same baby? Or: 'There was a discussion about rape in the staff room,' or 'It carried an important article about adultery by the Archbishop of Canterbury'.
- The non-sequitur can strike: 'Injured in the fighting in North Africa, Joe Bloggs retired last week.' The second clause bears no relation to the first. Even where the subject is clear there has to be a sequential connection for the sentence to be successful.
- If correcting the grammar of a sentence makes it sound stiff and pedantic it is better to re-write it until it sounds right.

Length

Because of narrow newspaper columns, short sentences (and paragraphs) are preferred in order to avoid an unbroken density of type. This does not mean that all sentences must be short, just that shortness is an advantage to the scanning eye.

Short sentences mean generally less complicated sentences with fewer clauses and less punctuation, so shortness can be an advantage in comprehension, too.

Short sentences are also faster and more graphic for the telling of hard news stories.

Here is an example from the *London Evening Standard*:

'POLICE fought a gun battle with an armed gang in an East London square today.

It happened as a team of police marksmen from the Central Robbery Squad were keeping watch on four men suspected of plotting an armed raid.

But the gang realized they were trapped and opened fire with shotguns. The police returned the fire.

Three shots were fired by police, but no one was hurt.

One of the officers, however, was taken to hospital suffering from shock.

A police motorcycle was damaged during the skirmish.

It happened soon after 9 a.m. in Carlton Square, Stepney.

Later police recovered three shotguns from the scene of the shooting.

Four men were this afternoon being questioned.'

This shows how short sentences and paragraphs (and also short words) give pace to a story. Of 117 words, 74 (or 63 per cent), are words of one syllable.

It would be boring, however, to have nothing but short sentences in news stories, and the subeditor must decide when longer sentences are justifiable or when a sentence should be split – sometimes with a little recasting – into two or more sentences. Here, from the *Birmingham Post*, is a story in which a slower pace is needed and into which longer sentences and paragraphs fit naturally:

'A Midland doctor investigating the outbreak of meningitis in Gloucestershire warned last night that an epidemic of the disease is likely to sweep Britain in the next few years.

His warning came as the world's leading expert on meningitis arrived in England to probe the outbreak which has killed two teenagers and affected 85 other people in the Stroud area of Gloucestershire.

Dr Carl Fresch, from Maryland in the US, is on a private visit, but will meet informally with doctors from Gloucestershire Health Authority who have been fighting the disease.

He will be working with another British expert on meningitis, Dr Dennis Jones, Director of Manchester's Public Health Laboratory.

Last night Dr Gareth Leyshon, Gloucestershire's district medical officer, said: "Stroud has been singled out for attention rather unfairly – there has been a rising incidence of meningitis across Britain, as in Norway, over the last two years, and we have not been affected more badly here than other areas."

Meningitis, which particularly affects children, is spread by coughs and sneezes, and in its most serious forms inflames the lining of the brain.

No vaccine has yet been found for the B type of meningitis discovered in the Stroud victims, but it is hoped that Dr Fresch may be able to identify more accurately the particular strain involved.'

Where the pace demands shorter sentences, or where the sentences are just unacceptably long, the subeditor should avoid the easy option of splitting a sentence by turning its clauses into sentences by beginning each one with an *and* or a *but*. Recasting will produce a better flow as well as better grammar.

It is acceptable in newspaper journalism to begin sentences with *and* and *but*, but it should be done sparingly to produce a particular effect of continuity within the context of sentence pauses, rather than as a device to break up a long sentence by replacing commas with full stops.

PARAGRAPHS

This is a thorny subject in newspapers. H. W. Fowler (of *Modern English Usage*) is right when he says: 'The paragraph is essentially a unit of thought, not of length; it must be homogeneous in subject-matter and sequential in treatment'. But he also says: 'The purpose of a paragraph is to give the reader a rest'.

With narrow newspaper columns, editors tend to give more stress to the second function, and paragraphing has come to be used to break up the text into readable nuggets more than to mark thought and fact sequences.

With the increased use of display in newspapers in recent decades came a reaction against the old style of columns of unbroken reading matter with few paragraphs. This has become exaggeratedly the case in the popular tabloids where it is assumed that readers are as allergic to long paragraphs as they are to long words. Taken too far, paragraphing, on this premise, can finish up identifying with sentences, however short, and as a result, balance and continuity of thought and fact are abandoned and the text is read in a series of quick jumps. Here, the plethora of indented and broken lines becomes as big a handicap to the scanning eye as solid text is.

The middle way between the desiderata of Fowler and the projection requirements of modern newspapers is to try to relate the visual breaks as closely as possible to breaks in the sense of the text. This is not usually difficult in the economic style of news writing common to most newspapers, though bad breaks will still occur if excessively long paragraphs are always to be avoided.

What is worse than breaking a paragraph in mid-thought is connecting two ill-matched thoughts or sentences into one paragraph, or ending a quotation and starting a description or a new quotation in the same paragraph. This can happen during editorial production when two paragraphs are run on together at page make-up stage so that a story carried in a number of 'legs' across the page can begin at the top of each leg with a full line and not a 'widow' or jack line.

The idea is to give a neat printed effect so that the eye does not turn to the top of the leg on to a broken line. Yet the yoking together of two unsequential paragraphs for this purpose frustrates the intention of the writer or subeditor, and damages the sense of the text.

As shown in the examples above, the length of paragraphs can be varied to suit the text. Action-based stories can be given pace through the use of short paragraphs as well as short sentences, while more leisurely stories such as accounts of ceremonials, royal occasions, functions, etc., or those involving explanations, can have longer paragraphs to give descriptive writing scope.

PUNCTUATION

Many ambiguities and misunderstood sentences are caused by faults in punctuation, particularly in the use of *commas*, either wrongly placed or removed. To say, 'The people who arrived because of the rain were accommodated in tents,' is not the same, for example, as saying, 'The people who arrived, because of the rain, were accommodated in tents'.

In the first sentence the arrival of the people was brought about by the rain. 'Because of the rain' is a defining clause. In the second sentence, 'because of the rain' is incidental. It is, therefore, called a commenting clause and is bounded by two commas. These commas make all the difference to the meaning of the sentence.

Take a sentence that might occur in a murder story: 'Jones killed his son because, he said, there was nothing to live for'. Then write it without the commas. Immediately the phrase 'there is nothing to live for' is transferred from Jones to his son. The meaning has totally changed.

Punctuation in newspaper texts is more concerned with clarifying meanings and avoiding ambiguities than in providing natural pauses, although the two purposes can coincide in longer sentences.

The old rule that commas should not precede the conjunctions 'and' and 'but' is generally worth keeping but the subeditor should be on guard for cases where the sense of a sentence requires a comma. In some cases 'but' introduces new information or a new thought and a comma pause can mark this helpfully as in: 'Yes, I think I ought to go, but I must be certain to get permission first'.

A comma is not needed in the sentences, 'It was not only the colour but the texture that turned her against it,' or 'He reached the top of the steps and turned to the left'.

Yet here is a case where the meaning hinges on the placing of a comma: 'They gave the prize to Jones, and his wife and the family were delighted'. Without the comma the reader would be at a loss to know whether Jones or Jones and his wife had won the prize.

Commas can be used to give stress to words and adverbial phrases, 'This was, evidently, the truth,' gives more stress to an important adverb than, 'This was evidently the truth'.

Commas are also used to separate items in a list, but their use to separate strings of adjectives is becoming less common and is not essential.

The best attitude towards commas is to use them only if, without them, the meaning is in doubt. This will avoid an excess of them. A sentence with many commas marking out many clauses is a potential source of misunderstanding and the subeditor would do well to split it.

The important thing is to know what the sentence is trying to say. If it contains an ambiguity because of bad use of commas (or for any other reason) that cannot be resolved by reference to the context, the subeditor should refer it back to the writer. A guess could lead to trouble.

There is no need of a comma between a house number and a street, as in 14 Coronation Street.

Full stop

Its use to indicate abbreviations as in Feb. and the Rev. is generally accepted, but not in abbreviations where the first and last letters of the full word appear, as in Cpl, Mr, Dr, etc. (For abbreviations see under 'house style'). A full stop is not used at the end of captions or in headlines.

Semi-colon

It denotes a pause longer than a comma but not as long as a full stop. In books and in feature writing there is still room for the semicolon by stylists who know how to use it and have more scope for polishing a sentence, but in news writing there is not a lot of call for it. It is sometimes used when a comma or full stop would have been adequate and its presence is often a sign that a sentence is getting too long. It serves a useful purpose in subdividing lists as in: '. . . east wing: three bedrooms, two with bathrooms; three dressing rooms, one with balcony; two staircases, one at either end, and a staff rest room . . .'.

Colon

This is a necessary but not often used punctuation mark in newspapers. It introduces lists of names, objects, qualities, etc., and precedes a quotation or explanation. It is pointless to use a colon followed by a dash for this purpose. Its use in a sentence to indicate a pause longer than a semi-colon but shorter than a full stop is now so rare, even outside newspapers, as to be extinct.

Dashes and brackets

Dashes are much overused and misused in newspapers on the assumption that they jolly up the text, and their rise in favour is in contrast to the decline in punctuation use generally. Here are some actual examples from newspapers of how dashes appeared:

> 'A bungling bandit robbed a bank – and left his cheque book behind.'
>
> 'But after Princess Diana left the Old Bailey at 2.10 p.m. – she did not follow tradition and visit the famous Number One court where a rape trial was in progress – her remarks were the talk of the courts.'
>
> 'Tour operators are hoping to make up for the disastrous summer season – one of the worst on record – by picking up extra winter bookings.'
>
> 'Three more were arrested at Blackpool – one of the seaside towns named in the IRA "hit list". The new arrests – they bring the total now detained to 16 – came after a police swoop at a house in James Gray Street, Glasgow – less than a mile from the tenement raided at the week-end.'

The first example shows the dash used to draw attention to the telling part of the sentence. This is the most justifiable use in these quoted.

In the second example the dashes are used to indicate a parenthesis, but the parenthesis is longer than the rest of the sentence and the result reads clumsily. The reader is also in doubt as whether the rape trial is part of the traditional fare at Number One court. This sentence should have been rewritten.

The third example shows another use of parenthetical dashes. Yet the effect is negated by tautology. Nothing would have been lost had the sentence said: 'Tour operators are hoping to make up for one of the worst summer seasons on record by picking up extra winter bookings'. The reader would have been saved the intrusive punctuation.

The worst misuse of dashes appears in the fourth example. Here the punctuation in two consecutive sentences consists of two full stops, a set of inverted commas, one comma and four dashes. The result is almost unreadable. The first dash should have been a comma. The second pair encloses an unnecessary parenthesis. The words could have read '. . . arrests, which bring the total now detained to 16, came . . .' without using any more words. The final dash in this sentence comes so quickly that the reader is in doubt which dash ends the parenthesis. Here a comma would have been sufficient, although of the four dashes used it is the one most justified by the sense.

Using dashes for commas is not only wrong, it is confusing to the eye, space-taking and devalues the dash as a significant text mark. It has resulted in the use of dashes being banned in some newspapers.

Using dashes to enclose a parenthesis is reasonable, though they are more space-taking than brackets. Grammarians are divided as to whether dashes or brackets signify the greater interruption. Whichever is used, the words contained within the parenthesis should be a comment or fact outside the flow of the sentence, but which the reader needs in order to digest its full import. The attitude of subeditors should be to stick to either dashes or brackets consistently for parentheses (many newspapers, in fact, shy off brackets, though they are shorter) but also to use parentheses sparingly. They impede the flow of the text and often indicate that the writer has not got his or her thoughts properly together.

Ellipsis

Another misused punctuation mark. Its correct use is to indicate that letters or words are missing, as in, 'According to the gamekeeper the man shouted, "You f . . . pig," and vanished'. Another use would be, 'There could have been a different ending had Helen only . . . but that is another story'.

It might be deduced from these examples that there is little use for ellipses in newspaper practice, yet they are becoming as pervasive as dashes. Here are some actual examples:

'A pony runs free through a meadow . . . dramatically reprieved from death at the eleventh hour.'
' "Oh Gawd . . ." said Geldof when I cornered him.'

A sports story in a popular tabloid adds to the confusion with these two examples:

'David Gower hit the jackpot at Trent Bridge yesterday . . . after working the "ten franc trick" on Australia in the third Cornhill Test. Allan didn't know what to call. There was no way of identifying heads or tails, so he shouted "francs" – and the other side came up.'

The first example seems to show an ellipsis being used in place of a dash to point a sentence, since there is clearly nothing missing from it. The second example could mean that words *are* missing, judging by the context. The sports example shows confusion reigning with both a dash and an ellipsis being used to point a sentence in the same story.

An even worse misuse occurs in some newspapers which use ellipses to enclose a parenthesis instead of dashes or brackets.

There are two points to be made here. One, it is inconsistent to use a dash *and* an ellipsis for pointing a sentence in stories in the same newspaper. In choosing one or the other for the sake of consistency the subeditor should bear in mind that an ellipsis is not suitable for this purpose, in any case, since it denotes missing letters or words, and also that the dash does the job perfectly well. Second, ellipses print badly on long print runs and sometimes the dots disappear from the page altogether, leaving mysterious blanks. Third, an ellipsis is the most space-taking of all punctuation marks and should be avoided for this reason unless being necessarily used for its correct purpose.

Exclamation mark

Sometimes called a screamer, the exclamation mark is beloved of headline writers and blurb writers who seem to believe it to be the carrot that attracts the donkey. There is a simple rule about an exclamation mark which should save subeditors a lot of trouble. It should be used only after an exclamation. It is unlikely that a phrase more than four or five words long could qualify as an exclamation by reason of the simple mechanics of delivery. Generally it is couched as a warning, a demand, a brief instruction or an ejaculation.

Beware of using a screamer to draw the attention of the reader to a non-exclamatory phrase or headline. Here are two examples from the same newspaper:

'. . . the terrifying thought flashed through his mind: "My God. I'm dead!" '
KINNOCK'S BUSTED FLUSH! (headline).

The first use is correct, the second without any justification. A truly justified third person exclamatory phrase is hard to achieve.

The over-use of exclamation marks in newspapers devalues them and looks silly and messy in headlines and text. Some cynical editors have been known to limit them to one to a page.

Apostrophe

These should be used *before the s* in singular possessives and *after the s* in plural possessives ending in s.

boy's   boys'

A common mistake is to fail to add an *s* after the apostrophe in proper noun possessives of names ending in s.

James's   Peebles's
Francis's   Lagos's

Note that the plural form of James and Jones and similar proper names adds *es* as in Jameses and Joneses.

Some well known place names containing possessives drop the apostrophe, for example St Albans and Earls Court. A gazetteer should be checked for these.

The apostrophe's original purpose of denoting missing letters is shown in its use in don't, won't, can't and it's (it is). In the case of possessive pronouns it has fallen out of use in his, hers, yours, theirs and ours, but is retained in one's.

There is no justification for using an apostrophe in plurals such as MP's, all the three's, or in expressions such as ten years' imprisonment or three weeks' leave, which are best regarded as adjectival phrases. On the whole, if there is doubt as to whether an apostrophe is needed, a subeditor should be guided by the sense. If the meaning is intact without it, leave it out.

Quotation marks

Inverted commas used to enclose quotations can be single or double but must be consistently one or the other. Most newspapers use double, which means they should be double in headlines too. The usual style is to have double inverted commas for main quotes, single ones for quotes within quotes, and double again for quotes within quotes within quotes (an extremity to avoid, if possible).

Quotation marks should be checked carefully on copy so that statements are properly attributed, and so that the reader knows when a quotation, once begun, ends. This is especially important in quotes within quotes. Main quotations should be introduced by a colon. For short ones in the run of a sentence or paragraph a comma is sufficient at the beginning and end. For example: The constable said he heard a voice say, 'Come quickly,' as he walked past the building.

Question mark

A fault to look for is a misplaced question mark in a sentence containing a quoted passage. If the quotation itself is a question, the question mark must come inside the closing inverted commas. If the question lies in the main sentence, then the question mark must come outside the inverted commas as in: 'Do you think it wrong to say, "No children will be admitted"?' The correct position can make a difference to the sense.

Hyphen

This much maligned punctuation mark that should be used sparingly but can sometimes be crucial to the sense. In line with the move towards less punctuation in newspapers it is discarded in familiar compounded words such as overuse, boyfriend, checklist, wartime, gasfilled, nosebag, handbag, and lowline, but kept in words such as co-operative, co-pilot, co-ordinate, and ill-used where its absence would confuse the eye.

Important uses to remember:

- To avoid ambiguity in certain adjectival phrases as between 'a model-manufacturing technique' and 'a model manufacturing technique', or 'a lost-business file' and 'a lost business file'.
- To indicate a compounded adjective as in 'an up-to-date method', or 'an away-from-it-all holiday,' but not, 'the method was up to date'.
- To distinguish different meanings of words as in re-formed and reformed, and in re-creation and recreation.

Hyphens, as with other punctuation marks, are discouraged in headlines to preserve visual neatness, but have an important use in typesetting systems to break words at the end of lines when justifying setting widths, or measures. It is not typographical style (or easy to read) to break a word in mid-syllable or to break proper nouns at all if avoidable.

Stresses

Accents are little used in newspaper printing systems to denote stressed syllables but italic is a common way of giving stress to a word beyond that contained in the punctuation. Italic was once used to denote foreign words but this is not now practised in newspapers where, in any case, most foreign words used have been acceptably anglicized.

7 SUBEDITING: THE WORDS ON THE PAGE

The two most used works of reference on a subeditor's table are an English dictionary and a gazetteer of place names, because of the need to get words and locations right. Only the foolhardy with however splendid a vocabulary would do without a dictionary since it gives not only spellings but meanings. It is important that the dictionary be the most up-to-date containing the very latest shifts and nuances in meaning or, better still, that there should be two dictionaries in order to compare definitions. Armed with these authorities, plus a lively and perceptive awareness of the spoken tongue, the subeditor is ready to embark on the editing process.

THE RIGHT WORD

The demands on space mean that there is no room in news stories for a long word where a shorter one can do the job as well. Ten long words excised in the space of a story three column inches long can make room for an extra sentence, which might allow the subeditor to bring in another useful fact (see 'headline words', in Chapter 8). Yet where words are substituted, care must be taken to ensure that the writer's meaning has not been changed or frustrated.

The right word is the one that is likely to be known to the reader and that makes the writer's meaning clear beyond doubt while at the same time being no longer than it has to be. A general guide is that foreign words should be avoided where there is an adequate English equivalent. It will be found that words of Anglo-Saxon origin such as house, bite, grip, flight, bold, sharp, bright and evil are shorter than their imported equivalent.

In news texts the subeditor would do well to avoid words that have more than one meaning, along with abstruse or academic words, circumlocutions and officialese of all types.

Technical words

Finance, computers, space weaponry, sociology – these are just some of the areas of science and commerce that have developed their own vocabularies. They are used and are acceptable in specialist publications but mean little to the general newspaper reader. Technical words should be introduced sparingly on news pages and only in contexts where their meanings are clear.

Yet in each of these areas, and in other similar ones, there are words that begin infiltrating the general vocabulary to win growing acceptance among readers. The space programme has provided lift-off, splashdown and hardware (a new use for an old word now meaning the machines and equipment). Economics has provided upturn, downturn and throughput; industry, blueprint, bottleneck and spin-off; the forces, bombshell, blockbuster and broadside. The examples could be multiplied. Image, model, reading, programme, strategy and target have all taken on new meanings through their use in technical fields.

Should such new words and new uses for old words be accepted? Certainly, if the context makes their use clear and they extend the meaning of the text for the reader better than any other word. Bottleneck, blueprint, blockbuster and spin-off, for instance, are colourful, metaphorical, almost 'visual' words that project an immediate mental image; but guard against applying them to situations where they are not justified. Is a plan really detailed and precise enough to be called a blueprint? Is a new book or film really a blockbuster in the effect it will have on the public?

Novel technical words used in general writing in a metaphorical sense are always in danger of 'catching on' to such an extent that they crop up all over and become clichés. So beware of over-use. Beware, also, of using too many at once. A piece of writing peppered with downturns, upturns, spin-offs, throughputs, lift-offs and 'go' situations has descended into jargon.

Foreign words

There is little point in using foreign words which have obvious English equivalents such as: rendezvous (arranged meeting), carte blanche (blank cheque, free hand), melee (mix-up, skirmish), cul-de-sac (blind alley, close), ad infinitum (indefinitely) and per annum (yearly). The following words of foreign origin are generally acceptable on the ground that they are not easy to substitute:

Ad lib	Cortege	Fiancé
Aide	Coupé	Fiancée
Aperitif	Crime passionel	Negligée

Attaché   Debut   Nuance
Blasé   De facto   Premiere
Bourgeois   De jure   Protegé
Brochure   Elite   Regime
Carafe   Entree   Repertoire
Cliché   Exposé   Status quo
Clientele   Facade   Sub judice
Corsage   Fait accompli   Venue

Circumlocutions

Journalists with any experience are not given to using long-winded phrases. Many of the circumlocutions listed in grammars are speech props such as 'as I stand here before you today,' or 'in this day and age,' or 'to all intents and purposes,' or are examples of officialese. A subeditor, while ever vigilant, would not normally expect to encounter such phrases in written copy, although some might enter via quotations, particularly from tape-recorded material. Some quaintness of phrasing is reasonable in quoted speech to preserve the flavour of the speaker's words but really space-wasting phrases should be removed, and usually can be without damage to the meaning.

Nevertheless, because of their lulling familiarity, some circumlocutions can tempt even subeditors into a trap. Here are some examples, showing the recommended use on the right:

Adjacent to   *Near*
Prior to   *Before*
As yet   *Yet*
As a result of   *Because*
In consequence of   *Because*
Currently   *Now*
At this moment   *Now, today*
As to whether   *Whether*
He is a man who   *He*
In order to   *To*
Tighten up   *Tighten*
Fill up   *Fill*
In the first instance   *First*
Owing to the fact that *Because*
Sound out   *Sound*
Check out   *Check*
Rest up   *Rest*
Try out   *Try*
Start up   *Start*

Meet up with	*Meet*
Meet with	*Meet*
Consult with	*Consult*
Inside of	*Inside*
He himself	*He*
Personally, I	*I*
All of	*All*
End result	*End* or *result*
At the back of	*Behind*
In front of	*Before*
At the side of	*Beside*
Join together	*Join*
In terms of	*As*
Acid test	*Test*
Each and every	*Each*
Extra special	*Special*
Face up to	*Face*
Horns of a dilemma	*Dilemma*
Out and about	*About*
True facts	*Facts*
Absolute truth, lies etc.	*Truth, lies* etc.

Synonyms

In order to avoid long words or repetition of words within a sentence or paragraph, or when seeking words for a headline with a limited type count, the subeditor searches for the alternative or shorter word which means the same. Beware here of the word that means *almost* the same but not quite. The main danger of this lies in headline writing (see Chapter 8) but meanings are at risk in the text too.

To say a man *claimed* or he *asserted* is not the same as saying he *said*. A change of verb for the way in which things are said can give undesirable colour or emotion to a speaker's words. To call a discussion or exchange of views a *row* invests it with a suggestion of violence. An *alibi* is not the same as an *excuse*.

Study not only the spelling and number of letters in the synonym you choose but also its precise meaning.

CLICHÉS

The term cliché is given to a wide range of hackneyed expressions, over-used phrases, tired adjectives, worn-out metaphors and current vogue words. Any word, or combination of words, if used excessively, it seems, is in danger of becoming a cliché.

Under this general umbrella are a lot of words and phrases that are unlikely to come the way of a subeditor since they remain in the limbo of their own environment. They are part of the jargon of their field, and consist of what, by newspaper standards, are tedious circumlocutions. Commerce has its own special ones: 'in this connection', 'for your information', 'it is considered that', 'I've said it before and I say it again', 'at the end of the day'. Everyday conversation has its own matching verbal props: 'as I was saying', 'if you see what I mean', 'I mean to say', and so on.

The danger of contamination from these sources is instilled into young journalists at the very start. What is sometimes not instilled into them is the danger from journalism's own shifting world of clichés. Keith Waterhouse, in his book *Daily Mirror Style*, has pointed out the changing fashion in newspaper clichés, some of them drawn from popular lore, others entirely invented by subeditors.

In the 1950s and 1960s the favourites were:

burning issue	dropped a clanger
cheer to the echo	and that's official
clutches of the law	the absolute gen
crying need	speculation was rife
fast and loose	monotonous regularity
red letter day	last but not least
just like the Blitz	out and about

By the 1980s a new raciness had begun to imbue clichés:

alive and well and . . .	pinta
Billy Bunters	cuppa
fashion stakes	taken to the cleaners
knickers in a twist	fairytale wedding
purr-fect (of cats)	sweet smell of success
clown prince	the end of the road
writing on the wall	sir (for teachers)
love child (nest etc.)	wait for it!
don't all rush!	to zero in
tears of joy	the name of the game
using your loaf	what it's all about

Commentators have a good chortle in pointing these out when writing about the press. Yet there is, in my view, a case to be made for the cliché in certain controlled circumstances. In writing for a popular readership, the journalist can use the newspaper cliché, like its relation the proverb, as a way of addressing the reader in familiar terms, provided:

1 It is not wastefully wordy.
2 It does not gloss or misrepresent the facts.
3 It is not used too frequently.

To refer to Billy Bunters in a piece about overweight schoolboys, or to a love child for a baby born out of wedlock, or to knickers in a twist in a funny story about charwomen, can strike a rapport with the readers of a popular daily where words like obese, bastard and muddle-headed might fail.

The term fairytale wedding, provided it does not exaggerate the nuptial splendour, conjures up an instant picture to readers which any other two words would find it hard to equal.

'The sweet smell of success' and 'the writing on the wall' can sum up particular human situations evocatively, provided they fairly reflect what is in the text.

In fact, there are clichés and clichés. It is the more specifically subeditor's gimmicks like 'Don't all rush!' or 'Wait for it!', purr-fect (in cat stories), pinta for milk and cuppa for tea that become most tedious by repetition.

The danger with clichés, in unthinking hands, is also that they can blur the facts of a situation by over-simplifying things for the reader, thus turning stories into stereotypes. They should be used only when they are apt within the context of the story and when they can make a point to the reader better than any other word or combination of words. In short, the odd cliché or two is acceptable, provided it earns its keep.

In using favourite metaphors, subeditors should beware of mixed ones, as in 'She was an angel of mercy pouring oil on troubled waters,' or 'He preferred to paddle his own canoe and cock a snook at authority'.

Over-used adjectives are particularly objectionable in a language as rich as English. Stunning, staggering, sexy, super, sizzling, terrific, luscious and amazing could be consigned unmourned to the waste bin along with superstar, megastar, zap! and phew!

Phew!

VOGUE WORDS

Writers on the use of English generally devote some space to what they scornfully call 'vogue words'. These are new words or new uses for words which have taken the fancy of speakers and writers and are being used a lot. It is difficult to generalize about them. Some, like *parameters* and *arguably* have a lulling polysyllabic charm for speakers with academic pretensions.

Others like *clinical, interface* and *syndrome* have a scientific ring about them which makes the speaker or user feel up to date. Some are useful and new ways of saying things without the need for lots of words and appeal to the busy. Some seem to have no justification at all.

What should the subeditor do about vogue words? The best advice is to accept them and use them where they help to clarify a meaning or communicate a point to the reader but not to overuse them. English is so rich and flexible a language that no word need be overused. Rather should one be looking for nuances of meaning that new words bring with them, thus accepting them as a broadening of the language rather than a restricting of it. One should be certain and united, however, about what they mean. If there is any doubt or ambiguity about their meaning, then they should be left alone until the grammarians have sorted them out.

Here are some examples of current vogue words, one or two of which will be found usefully deployed in the text of this book. They should be treated on their merits. Some of them could be the Queen's English of tomorrow.

Accord	Agreement. Began popular life as a headline variant. Not much justification for it in the text.
Aggravate	Has subtly added to its meaning to annoy, or cause trouble as well as to make worse. The usage dates back to the vogue of its abbreviated form aggro during the Teddy boy period in the 1950s and 1960s – hard to reverse a trend like this, but it is best left to quoted speech.
Ambience	Once an encompassing circle or sphere, now aura or atmosphere. Useful for descriptive writers but not a good word in a hard news story.
Arguably	'Capable of being argued as,' but mostly wrongly used as 'more than likely to be,' or 'almost certain to be'. A perfectly good word which is simply getting too much to do and deserves a rest.
Axiomatic	Used to be used only in scientific proofs; now in vogue as conferring certainty on a statement or conclusion on the lines of 'it goes without saying'. A silly misuse.
Basically	A tiresome prop with which to begin an explanation. A very expendable word.
Charisma	Formerly a special or God-given grace or talent. More recently a special quality or aura displayed by someone. An overused word, but there is no other word that says precisely this. Use sparingly.
Chauvinism	Exaggerated and bellicose patriotism; now adapted to mean glorification of the sex – e.g. male chauvinism – and seems to have taken root. It is doing no harm used with the

	word 'male' and describes an attitude that is recognizable.
Clinical	Pertaining to the sick-bed (as of clinics). Has mysteriously come to mean *coldly* and *detachedly* (as in an action). Best avoided.
Concept	Used a lot where the word *idea* used to be used. Has a pseudo-scientific ring. Use idea – it is shorter and is mostly what the writer means.
Contact	A useful noun-verb when one is not certain of means or method. Otherwise say *meet, telephone, write* to or *call on*.
Criteria	Necessary requirements upon which a judgement or decision is based. There seems to be a lot of criteria about these days but the word mostly does a good job.
Dialogue	A vogue word from the international political scene. Strictly it is between two people and is wrongly used for discussions, meetings, talks, etc., and could do with being aired less.
Ecology	Is beginning to take over from *environment*, meaning the natural world in which we live. It is stretched to mean anything which is natural as opposed to man-made, though originally it meant simply the branch of biology which dealt with the relationship of plants and animals to their surroundings. No stopping this one.
Escalation	Another bastard from the international political or war scene. Has developed in recent decades as a back formation from *escalator*, a moving staircase. It means to increase or develop by successive stages. The world's trouble spots are keeping it well employed and it has carved a patch for itself.
Fruition	Plans everywhere are coming to fruition, yet the word has nothing to do with fruit or bearing fruit, but means the act of enjoyment or pleasurable possession. Often wrongly used.
Hopefully	Means with hope, but is now worked to death meaning 'it is hoped that . . .' or 'I hope so'. Should be given a rest.
Image	Not so much a copy of an original as a special, highly glossed version to present to the public; more a façade than a copy. But everyone who is anyone now seems to have one – or need one.
Interface	A surface separating two portions of matter or space. Became in the 1960s a region or piece of equipment where interaction occurs between two systems. More loosely it has come to mean any form of joining together. Treat with caution.
Line	A particular type of argument or set of explanations, e.g. the Marxist–Leninist line, or the line adopted. *Argument* is more descriptive, and preferred, despite being longer.

Maximize	Meaning to work or pursue something to the maximum degree possible or feasible, has come to stay. Although one of the prolific new 'ize' verbs it is pithy and colourful and says it in a word.
Meaningful	A much used invention of recent decades meaning full or replete with meaning. I can't bring myself to condemn it since it does a job – but beware of overuse.
Minimize	Handy – as with *maximize*.
Mix	A useful adman's word meaning the sum total of all the various ingredients. An insinuatingly useful word which has the extreme advantage of brevity.
Normalize	Means to cause to return to normal, akin to regularize. These are handy verbs, though too many of them on a page can cast a blight.
Ongoing	Very popular for *continuing* or *never-ending*. Has its uses but is often used unnecessarily like a verbal prop.
Parameters	A scientific word meaning qualities or factors, which has achieved runaway vogue as a synonym for limits or boundaries. A snob word that has caught on. There seems to be no need for it in the context in which it is usually used.
Prestigious	Having or manifesting prestige (it formerly meant practising juggling or cheating). It is now cheerfully bestowed upon people, jobs, property, businesses and sites. If only it were used a little less . . . though it is clear there is a lot of prestige about.
Proliferation	The endless development, formation and spread of things, is a useful word which does a job. One can live with its popularity.
Scenario	A script-writer's word which describes something more detailed than just the scene. Is sometimes the right word, but is too exotic to stand much use.
Situation	We are surrounded by situations. Every decision, development, spending programme, injection of funds, order and cancellation depends on a situation. There are go situations, stop situations, study situations, new situations, bad situations, ongoing situations – too many of every sort of situation except situations vacant. Blame the politicians and the economists.
Spectrum	A pseudo scientific word for *range*. Use *range*.
Symbiosis	A withdrawn cult word used a lot by those who think they know what it means. Strictly: living or involved together in mutual support. One for newspapers to leave alone.
Syndrome	A set of concurrent things or symptoms, now used for many

	conditions of the body, mind and imagination. Is becoming generally accepted and puts things in a nutshell, even if there are more syndromes about than we thought possible.
Thematic	Having or pertaining to a theme (of art forms, philosophies, systems and instructions). A useful descriptive word that earns its keep.
Thrust	As of an argument or explanation. A nice phallic word much favoured by aware males. Does a job.
To host	Meaning to preside over an invited gathering (of friends, visiting heads of state, teenage delegates, etc.). It does a useful job.
Traumatic	Which once meant pertaining to, or caused by, injury or shock, has become almost a synonym for dramatic. Though it hardly deserves to, it sounds good.
Trendy	Used for things that are going to be 'the thing' or that were 'the thing' and are no longer so. A word to bury.
Update	As a verb or noun is a useful technical addition to the general vocabulary. It is brief and precise and no other word seems to mean this.

MISUSED WORDS

A more general danger is the misuse of some words in print because of popular misconceptions about them that have stuck and evaded the writer and subeditor. For instance *chronic*, in the case of illness, means lingering and not necessarily severe. Buildings that catch fire are not *razed* to the ground but simply razed. People are not *exonerated* from blame, but simply exonerated; while in spending programmes and budgets, *targets* are there to be hit, not exceeded, and *ceilings* reached, not surpassed.

Here is a list of words commonly misused or misunderstood which subeditors should watch for:

Alibi	Latin for *elsewhere*. To offer an alibi means to explain that one was elsewhere at the time. Not to be used for *excuse* in general.
All right	Spell as two words, not *alright*, except in slang text.
Alternative	Every other. Not to be confused with *alternatively*, offering one of two possibilities.
Among	Used where there are more than two people or things. Not to be confused with *between*, which distinguishes between two people or things.

Subediting: the words on the page 129

Anticipate	Should not be used for *expect*.
Appraise	To form a judgement about something. Not to be confused with *apprise*, to inform.
Apprise	See *Appraise*.
Avoid	To have nothing to do with. Not to be confused with *avert*, to turn away from. You *avert* trouble by *avoiding* its likely cause.
Beg the question	Not to evade a straight answer, but to give an answer based on an unproved or unacceptable assumption.
Between	See *Among*.
Can	Not to be used for *may*. Can means able to, *may* means allowed to or permitted to, or has a chance or possibility of doing. Differentiate between 'you can do it,' and 'you may do it'.
Claimed	Not an accurate substitute for *said*, *stated* or *declared*. It suggests an element of dispute about what is being said or stated.
Compare to	To liken one thing to another.
Compare with	To note resemblances and differences between two things.
Consensus	Not *concensus*.
Credence	Means belief or trust. *Credibility* is the quality of being believable. *Credulity* is readiness to believe.
Credibility	See *Credence*.
Credulity	See *Credence*.
Declared	More precise and emphatic than *said*. See *Claimed*.
Different	From, not to or than.
Dilemma	A choice between two alternatives. Not to be used in general for difficulty of choice, or for weighing up a situation or problem.
Discomfit	Means to overwhelm, to defeat, to disconcert. It has nothing to do with *discomfort*.
Disinterested	Means not tempered by personal interest. Do not confuse with *uninterested*, without interest in.
Dissociate	Preferable to *disassociate*, not to be associated with.
Due to	Should not be used in place of *because*, as in, 'he was delayed due to the weather'. Leave *due to* phrases including 'the respect due to . . . etc.'.
Economic	Mostly now associated with the economy
Economical	Relating to economics.
Farther	Of distance (*see Further*).
Fewer than	Of numbers. Do not confuse with *less than*, of quantity.
First	Say, *first, second, third*, not *firstly, secondly, thirdly*, etc.

Fix	Vague as a verb. Where possible use *arrange, attach, organize,* set up, etc.
Following	Do not use instead of *after* or as a *result* of. It is less precise.
Forensic	Simply means pertaining to words of law – nothing more.
Forgo	Is to abstain from. Do not confuse with *forego,* to precede.
Forego	See *Forgo.*
Forward	Not usually *forwards.* The *s* is mostly dropped from *homeward, backward,* and *sideward,* but is retained in *towards.*
Further	Of time or degree (see *Farther*).
Hanged	Criminals are *hanged.* Things are *hung.*
Happened	Without warning, as with *occurred.* Took place suggests planning or forewarning.
Historic	Part of history.
Historical	Concerned with historical events and records.
i.e.	From the Latin *id est* (that is), and should be used to introduce a definition. E.g. (*exempli gratia*: for the sake of example) should be used to adduce an example.
Imply	To suggest without stating directly. Do not confuse with *infer,* to deduce or draw a conclusion from.
Infer	See *Imply.*
Insure	To provide for damages or replacement in the event of loss. Do not confuse with *ensure,* to make certain, to guarantee.
Last	Of more than two things.
Latter	Of two things.
Leading question	Not a question that is difficult to answer, but a question that is so designed that the answer is suggested.
Lend	More correct than to *loan,* which is an obsolete old English verb which has come back to enjoy a vogue in the US.
Lengthy	*Long* is mostly better (and shorter).
Less than	See *Fewer than.*
Leave	Not to be misused for *let.* Say 'let it remain as it is'.
Loth, Loath	This adjective, meaning unwilling, is correct in both spellings but is best spelt without the *a* to differentiate it from to *loathe,* meaning to detest.
Militate	See *Mitigate.*
Mitigate	To make less severe or serious. Do not confuse with *militate,* to act as a strong influence.

Nice	Too vague now that it has lost its old meaning of fine, balanced. Use a more precise adjective.
Practical	Useful in practice.
Practicable	Capable of being carried out.
Practically	Has come to mean *virtually*. Better to use *virtually*, which has only one meaning, or *almost*, which is shorter and simpler.
Protagonist	Means advocate or champion and not necessarily the opposite of *antagonist*.
Refute	Means to prove wrong, not to deny or repudiate.
Resource, recourse, resort	A source of much muddle. *Resource* is a stock or reserve to draw upon. To have *recourse* is to return to or to fall back upon (one's *resources*). *Resort*, in this area of meanings, is that place or thing or person upon which one depends for a solution. 'A royal pardon was his last resort.'
Respective	Pertaining to those in question is the adjective and *respectively* the adverb. 'The *respective* authors,' but 'The authors were, *respectively*, etc.'
Seasonable	Suitable for the season or time of the year.
Seasonal	Occurring in association with a particular season. 'Hot weather is seasonable in the summer,' but 'Most holiday work is seasonal'.
Stated	More fully covered than *said* (as of a statement), not so emphatic as *declared*.
Took place	See *Happened*.
Transpire	Became known, not *happened* or *occurred*.
Try	Try *to*, not try *and*.
Under way	Beginning to move, two words. Formerly *under weigh* (anchor).
Unique	Avoid *quite*, *most* or *rather* unique. There can be no degrees in uniqueness.
Verdict	Finding of a jury. The judge gives his decision and sentence.
Whence	'She returned whence she came' (not *from* whence).
Wise	Tax-wise, price-wise, etc. are mostly and preferably avoidable.

HOUSE STYLE

In preparing copy for typesetting, subeditors have to be familiar with the house style of the newspaper in order to avoid textual inconsistency in spelling, numeration, the use of abbreviations, the Anglicizing of foreign words and in other areas where alternative uses exist.

Most newspapers, having opted for particular ways of doing things, guard their house style as zealously as their typographical style and can come down heavily on its neglect. Subeditors are expected to prevent style faults getting into the paper.

The justification for sticking rigidly to house style is not just that the newspaper's version is necessarily the right one but that spellings, abbreviations and also style in numeration are confusing to the reader, untidy, and also unprofessional, if rendered in different ways on the same page. An imposed consistency is the way out.

Some style problems are resolved by adopting recommended uses in such works as Collins's *Authors' and Printers' Dictionary* or *Hart's Rules for Compositors and Readers*. Some newspapers simply opt for the shorter or more contemporary or more 'English' of any given alternatives, while others go for etymological exactitude (with its risk of pedantry) or simply follow a style because it has always been the style.

Whatever the reason, here are some areas where new subeditors should reach for the house style sheet when in doubt:

Spellings

Inquire / Enquire — Authorities who try to prove the words mean different things are nit-picking.

Gaol / Jail — The old Norman French legal word is giving way to the more modern 'jail'.

Connection / Connexion — The new 'connection' is ousting the old English form.

Despatch / Dispatch — Commoner with the 'i'.

Marquis / Marquess — 'uis' preferred.

Gipsy / Gypsy — Both forms remain popular.

Judgment / Judgement — The 'e' is losing ground.

Swop / Swap — The 'a' version is growing in favour.

Transatlantic   A medial cap *A* is seldom used.

The use of 'ize' endings in such words as organize and nationalize as against the 'ise' for merchandise, advertise, etc., (despite the advocacy of the *Oxford English Dictionary* and the *Authors' and Printers' Dictionary*) is losing ground and 'ise' is appearing in all cases. The differentiation was never popular in newspapers. It is still used in publishing houses – and in the text of this book.

American words

The use of American spelling is generally discouraged in British newspapers even in quotes from documents. Subeditors, in American-originated copy, should see that railroad is railway, that a car fender becomes a bumper, and that specialty becomes speciality. Beware the use of the word subway. Color, honor, rigor, glamor, etc. should have their 'u' restored and defense should read defence.

Yet peddler is often used in place of the English pedlar where drug peddling is the subject, and program allowed to stand in computer contexts.

Anglicized words

Wide variation exists in the spelling of foreign names, with Russian composers and Chinese cities turning up in particularly bizarre forms. Tchaikovsky can be rendered Chaikovski and Scriabin Skyryabin. Peking can be Pekin, or even Beijing, while Hangkow can turn up as Quanzhou.

More simply, Capetown and Hongkong are alternatively Cape Town and Hong Kong. Then there are Jugoslavia (Yugoslavia), Baghdad (Bagdad), Iraq (Irak), Tokyo (Tokio), Tehran (Teheran), Khartum (Khartoum), Bucharest (Bucarest) and Rumania (Romania, Roumania).

Capital letters

In the eighteenth and even the nineteenth century capital letters littered writing to denote qualities, word stress, people's titles and specific references to things. Now they are the exception. The Prince or the Duke is used where a specific person has been introduced by name and continues to be referred to and where the title is part of the person's name, but a capital is not used where the word is simply the rank or appointment – judge, chairperson, secretary, etc. A country's President takes a capital, however. Some papers use capitals when introducing an important official for the first time, as in 'Trevor Griffiths, General Secretary of the National Union of Teachers,' thereafter referring to him as Mr Griffiths. The Royal Family is capitalized but other uses of royal tend to be lower case.

The seasons can be capitalized or lower case according to style, specific areas like the North, the West Country, etc., capitalized but points of the compass more frequently in lower case. The Government takes a capital but government in general lower case. Left-wing and Right-wing, or simply the Left and the Right, are usually capitalized to show a political meaning.

Capitals used in abbreviations are appearing more and more without points, as in USSR, UK, EEC, IBM, NCB, etc, but U.S. takes points so that it does not mean 'us'. Acronyms (combinations of initials pronounced as words) take no points and are often given in lower case after an initial capital as in Unesco, Mensa and Nato.

Generally, the use of capitals in preparing copy is lessening, lower case being sufficient unless special significance requires a capital. In style sheets the aim is consistency, one way or the other.

Names

The usual style is 'John Jones, aged 20,' but sometimes 'aged' is missed out. Street names are occasionally hyphenated as in '25 Church-street, Norwich'. People are referred to after the first mention as Mr, Mrs, Miss, Ms, except in criminal court cases where the style is usually by surname only.

Numbers

Most style sheets give one to ten in letters and eleven upwards in numbers. Fractions can be an exception, $5\frac{1}{2}$ looking better than five-and-a-half. Figures are always used for mathematical formulae and percentages. Sentences should never begin with a number given as figures: 'Three hundred years ago, a leading poet . . . etc,' is better (see Chapter 8).

With dates, many style sheets go for the logical day, month, year method – 23 February, 1987, but some prefer the month first. Inclusive dates are best rendered 1986–87, 1915–16, rather than changing just the last digit.

With money, amounts are mostly rounded down in headlines, £509 becoming £500 and with an 'm' used for million as in £35m. In copy, £35 million is mostly preferred to £35,000,000, with figures being used for more precise amounts. In smaller sums, say £49.07, but 49p, using 'p'.

Bans

Look out for banned words especially diminutives such as kiddie, hubby and doggy. Every editor has his or her phobias and will sometimes put a much used headline word under total interdict.

Cyphers

The ampersand (&) is little used in newspapers despite its brevity. Accents are mostly missed out, the dollar sign not always in the type range, and the % percentage sign tending more and more to be written as p.c. or per cent.

Abbreviations

Most letter abbreviations are written without points (see under 'punctuation' above) with acronyms in lower case after the initial capital. In the case of common abbreviated titles as in Prof., Mr and Dr, many style sheets leave off the point if the last letter of the word is included. Headline abbreviations carry no points unless the meaning is in doubt without them. Organizations should be given fully at first mention so that the abbreviation thereafter is understood. In long texts it is useful to repeat the full name, even if in a shortened form, to refresh the reader's memory.

With ranks there is wide divergence, as in Lieut-Cdr (Lt-Cdr), Flt-Lieut (F/Lt), Lt-Col (Lieut-Col), Constable (PC), Corpl (Cpl) etc. The style sheet has to be studied. The use of county abbreviations such as Beds, Berks, Bucks,, Oxon and Salop should also be checked, as should abbreviations for the longer months such as Aug., Sept. and Nov.

Weights and measurements, where giving precise amounts, are usually abbreviated thus: 10 kg, 4 lb, 6 fl oz, and are not pluralized with a final 's' or given a full point. If written out in full, as in 5 kilograms, the normal plural 's' would apply.

Excessive use of a variety of abbreviations in the text is tiresome to the reader and should be avoided. The space saved is often more than counterbalanced by the loss in clarity.

Avoid using time-saving abbreviations in copy for keyboarding in case aftn (afternoon), btwen (between), yesty (yesterday), mng (morning) or chmn (chairman) turn up like that in the paper.

Typographical style

It is important to be consistent in the text in such things as film and book titles, quoted verse, the names of newspapers and popular songs and the style and title of MPs, church dignitaries, etc. Some newspapers give all titles of songs, books and even newspapers in italic. Others give them simply with initial caps. Song titles are sometimes quoted. The use of quoted verse should be checked with the style card. MPs can be given as John Jones (Con. Glamorgan) in a political report, but as John Jones, Conservative MP for

Glamorgan in other contexts. Style and titles of church dignitaries should be carefully checked in reference works if not given in the style sheet.

Each newspaper has its regular style for setting tabulated work such as television programmes, racing cards and election results which should be followed. Look out for formatted setting codes for use with VDUs.

Trade names

Most firms take umbrage if the trade names of their products are used as generic names without capital letters. The angry letters received on this subject have resulted in all newspaper offices having lists of trade names with their equivalents either as part of house style or listed separately. These have to be learned. Examples include:

Hoover	vacuum cleaner
Kleenex	paper tissues
Fibreglass	glass fibre
Thermos flask	vacuum flask
Biro	ball-point pen
Elastoplast	tape dressing
Oxo	beef cubes
Nylon Terylene	artificial (man-made) fibre
Martini	vermouth

8 HEADLINES: CONTENT AND APPROACH

Headline writing is one of the more difficult subediting jobs. It takes high concentration on the materials of the story – the facts and supporting quotations – to render them quickly into a few short words that will tempt the reader to read on.

Some subeditors have a natural flair for this at almost the first reading of the copy but, mostly, headline writing comes with painstaking practice. The best are not necessarily the quick off-the-top-of-the-head ideas but can be the result of a patient juggling of words on a copy pad after the subbing, on 'hard' copy or screen, is completed. Here, the morning paper routine, particularly on a well staffed national paper with its longer time factor, is an advantage. There might be time for discussion in the case of a big story, and the most effective headline can be a 'committee job' in which improvements have been worked on the ideas of others. There is, thus, every reason why a national paper should show polish in its headlines compared with the town evening with its tight edition deadlining.

Yet there is no room for sloppy ineffective headline writing on any newspaper if it is to succeed in getting and holding the reader's attention.

It is useful, in considering the technique, if we start with a definition. We can say that the headline has two main functions:

1   It draws the attention of the reader to the contents of the story.
2   It forms part of the visual pattern of the page.

The first function has to do with words. The subeditor is the synthesizer, filtering material so that its essence is refined into a simple 'read me' message. This must be achieved without doing damage to the facts by over simplification or 'bending'. In other words, under the pressure to achieve a good headline, the subeditor must guard against distortion.

Here, the subeditor comes up against the second function of the headline. For the technique of modern newspaper layout demands that some sort of type pattern be imposed on newspaper pages. On news pages, stories are allocated a size of headline type and width of setting to suit the story's relationship to the page as a whole. This limitation nominates the maximum number of letters, or characters, that are possible in each line. In synthesizing the story in terms of a headline the subeditor (except in rare cases where the layout can be changed) has to accommodate the words to the letter count that the type allows.

THE WORDS

To attract the eye a good headline must have *clarity* and *impact*. Clarity means that the words convey to the reader what the subeditor intends without confusion or ambiguity. Impact means that the effect of the words is strong enough to persuade the reader to read on.

It will be seen in the pages that follow, that headline language, in its simplicity and immediacy, is close to everyday language – the demotic spoken tongue of a country with its familiar, generally short, words. Yet in its construction, a headline is nothing like recorded speech, for to say what has to be said it has to be as deliberately composed as a metrical poem. It is language pared down to the bone, producing the maximum effect from the minimum of words which, in addition, must submit to a typographical discipline quite alien to ordinary speech. Every word has to be carefully weighed and measured, key words located and not lost, legality and factual accuracy considered, and a visual effect achieved with the result.

Subject

Since it is, in effect, a condensed sentence, the headline has to have a subject and a verb in the right place. The subject is what the headline is about. THE QUEEN ABDICATES would be a blockbuster example of both subject and verb at work. FIVE HUNDRED DIE IN JUMBO JET CRASH shows that it is the deaths of 500 people rather than the crash itself that is the subject. The subject comes first in an effective news headline. It is the essence of the subeditor's summation of the material of the story. It is what the story is about.

While, in the text, an implied subject can sound right in a context that makes clear what it is, a subjectless heading should be avoided. I read in my local paper the headline BLAMED DRINK, and in an American paper STEALS PISTOL TO END LIFE. This is headline space badly used. There are better and clearer and more compelling ways of attracting the reader's attention.

Verb

The active voice is stronger and uses fewer words than the passive voice. STORM POUNDS BEACHES has more pace and immediacy than BEACHES ARE POUNDED BY STORM. BOY MEETS GIRL is better than GIRL IS MET BY BOY. But beware of distorting the headline in trying to avoid the passive word. ROCK STAR IS KILLED IN SKI FALL is better than the more active SKI FALL KILLS ROCK STAR because it has the subject in the right place. A headline that relegates the subject to the last two words is a weak headline.

Yet the verb must be used as soon as possible. It is the verb that energizes the headline and gives it pace, compared to which the adjective, for instance, has a modest role.

Special words

Whatever the words, the subeditor should go for plain English. Polysyllabic or unusual words are a put-off to the reader except in newspapers in a specialist field. They also rob a headline of pace as well as impact.

At the same time the subeditor should try not to create a special headline language by using words that only exist in terms of their headline use (see Chapter 9, 'Alternative words').

Omission of words

A number of ploys are used by subeditors to give a news headline pace while containing its message within a given type and measure. Some of these, if not used carefully, can endanger the clarity of the words.

For example, the auxiliary verbs *to be* and *to have*, and some others, can be omitted where they are implied beyond all doubt in the wording. SIEGE GUNMAN GIVEN LAST WARNING sounds right without the *is* after gunman. But JOBS PLAN HOPE FOR WORKLESS should include either *is* or *give* after hope to avoid ambiguity. Plan could strike the reader at first as a verb and not a noun. SICK PAY UP is tantalizingly ambiguous without the *is*.

LAB FIRES PROBE BY CID boggles the eye until we learn from the text that it means LAB FIRES *are* PROBED BY CID.

ABUSES INQUIRY IN S AFRICA 'INADEQUATE' tempts us into thinking there is an implied subject with *abuses* as its verb. But *abuses*, it turns out, is a noun-adjective describing inquiry. The auxiliary *is* after Africa would have made this clear.

Nouns as adjectives

Nouns as adjectives, sometimes compounded, are a useful short cut in identifying the subject without using too many words. *Abuses inquiry*, in the headline above, is shorter than *inquiry into abuses*, but a headline like FAN CLUB FIGHT DRAMA is baffling until we realize that it is a label headline and that *fight* is not the verb we are looking for but the subject and that *fan club* is two nouns compounded into a noun-adjective.

Likewise, in JUMBO CHECKS ORDER we discover on reading the story (if we are tempted to read it by such a muddle of words) that *jumbo* is the noun-adjective (jumbo jet) and not the subject, and that *checks* is the noun and not the verb, and that the headline is a verbless label telling us that a check on jumbo jets has been ordered.

These faults show the ambiguity and obscurity that lurk when nouns are used as adjectives in headlines that are not activated by a verb. In all the examples quoted above (which really *were* used in newspapers) the faults might have been detected had the subeditor mouthed the words half aloud. A good headline should sound good as well as reading right.

Here is a headline in which the subject has a compounded adjective made up of three nouns and a verbal phrase (not a recommended idea) but which is saved by an active verb following the subject, as it correctly should:

JET CRASH GIRL'S 'WAKE UP' PLEA SAVED MOTHER

Let us look at one more actual example of nouns used as adjectives. MYSTERY MAN FACES DEAD-PRIEST CHARGE shows an adjective-noun phrase being used as an adjective to describe the criminal charge in a headline in which the subeditor is trying, for legal reasons, to avoid using the word *murder*. The effect is acceptable and produces a headline that is clear. It also tempts us to read on to find out more.

SYMBOLS

A metaphor or phrase that can sum up a complicated situation and signal it clearly to the reader is a subeditor's dream. It not only eases the job of containing the message in the type but can give a colourful headline.

PRINCE WOOLLIE COMES HOME, from *The Sun*, not only symbolizes a picture of the young Prince William swathed in woollies on a cold summer's day but is a clever play on the name.

In its different market, the *Financial Times* is not averse to using metaphor and literary allusion to create a symbol headline in an unlikely setting as in:

DANISH ENGINEERS
TAKE A TILT AT
WINDMILL MARKET

Sport is the great area for metaphor, with this cricket headline in *The Guardian*, ENGLAND SAVOUR LEG OF LAMB (about a cricketer named Lamb), an extreme example.

The more everyday use of lovenest, lovepact, sex-bomb, whizz-kid, brain-drain and latch-key babies as situation symbols shows the extent to which headline imagery has filtered into the language. The danger at this level is that these identity flags will lose their potency through overuse. Also in this danger zone are such verb-object expressions as 'gets the go-ahead', 'fires a broadside', 'the writing on the wall', and 'blows the whistle on'.

Word accuracy

Since a headline gives a story greater prominence than does the text, the subeditor, as we said, must be careful not to damage the accuracy of the facts in the search for a compelling headline.

Emotive words like hysterical, failed, regretful, ruthless, resentful, glib, cruel, etc. should be used with care and only if they are supported by evidence given in court or by a magistrate's or judge's summing-up. Verbs such as critical, attacked, condemned and scorned must be borne out in the text by attributed quotations. People's words should not be pushed beyond their meaning to stand up a headline.

Beware of devaluing words such as slashed, slammed, lashed, smashed, rushed, raced, grabbed, slaughtered and demolished by using them to pep up headlines when the material of the story does not match up. The sports pages are the biggest offenders here. In one issue of a national tabloid the following headlines appeared: TELFER LASHES LIONS, POLE-AXE FOR POOR BUSTER, INTIKHAB BLASTS A WARNING, JACK FLATTENS SUSSEX CRICKET, WILLIS CALLS FOR KILLER TOUCH, and SOBBING SUE SMASHED. Violence through exaggeration in the sports pages can make violence on the field seem like small beer.

HEADLINE PUNCTUATION

A headline is a condensed sentence so as well as needing a subject and a verb it also needs punctuation. A quick examination of any headline, however, will show that commas, dashes, question marks, etc. look untidy in big type.

While keeping in essential punctuation, it is a good idea to write headlines that need little or none. The following points will help.

Full points are not used now except in long free-style rhetorical headlines (mostly on features pages) which have a sentence break as in:

> GO TO WORK ON AN EGG
> THEY SAY. NOW WHAT
> ABOUT THIS FOR A
> ZANY BREAKFAST IDEA?

A *colon* is a more useful break in a two-idea headline, as in:

> MID-EAST WAR
> LOOMS: SYRIA
> MOVES TANKS

But even the colon is being phased out on many papers. Instead, a second headline idea, if wanted, is given a deck on its own. A colon can be a space saver in place of a dash as an acknowledgement, as in I'M DOING IT MY WAY: SINATRA IN LONDON TODAY.

Commas are sometimes necessary but should be avoided if possible. They can be used to note a missing *and* as in NEW INVESTMENT BOOM IN COINS, STAMPS. A comma is not necessary before an attribution, as in MY WIFE THUMPED ME SAYS MAN. If a comma is necessary to avoid ambiguity then the headline should be rewritten.

Question marks should be avoided. A newspaper should not be asking the reader things. It is supposed to know. An exception would be in a *dialogue* headline, such as BANISHED? NOT ME SAYS HUSBAND.

Exclamation marks (or screamers) should be restricted to exclamatory phrases like Howzat! Cheers! or Knocked out! and are most suited to sports pages. They are pointless as used in the following actual tabloid paper headlines:

> £85,000 JET-SET CAR IS A LOCK-OUT! (which also labours under figures, a pound sign and two hyphens).
> OLD MOORE SEES 10 YEARS IN THE DOLDRUMS! (an eight-word exclamation).

Quotation marks are best used single in headlines. Apart from enclosing actual quotations they can be used around words to indicate doubt or an unconfirmed assertion as in: 'DEATH TRAP' CAR BLAMED FOR CRASH, or to show a fact to be false as in 'DEAD' MAN IS WEDDING GUEST. Single quotes also get over a problem where two or more words are compounded into an adjectival noun as in: 'GROW YOUR OWN' CAMPAIGN FALTERING.

They are not enough to sustain on their own an uncredited quotation.

Either there must be a *tag-line*, or the person's name should be included in the headline or its second deck.

Ellipses are strictly dots to denote missing letters or words, but they are used in newspaper headlines for a variety of purposes from replacing a colon to filling out short lines. They are best banned along with the exclamation mark as the example in Figure 27, which actually appeared, shows.

GIVING BIRTH TO A ROYAL MYTH..!

Figure 27 *Punctuation that a headline can do without*

HEADLINE ABBREVIATIONS

Abbreviations, especially those formed from initial capitals, make a headline obscure to some readers. A large number do not know what UNESCO, NATO and EEC mean. TV, BBC and the TUC pass muster but unions with such names as the Electrical, Electronic, Telecommunication and Plumbing Union; the Amalgamated Union of Engineering Workers, and even the National Union of Mineworkers are better referred to in headlines as the electricians, the engineers and the miners rather than by their initials, the EETPU, the AUEW and the NUM. The full name can be explained in the text.

Capital letter abbreviations look ugly when they do not constitute a word in the mind's eye – an acronym – and are best rationed to one set to a headline, if they must be used at all.

Dr, Mr, Coun., Capt., the Rev. and Prof. are inescapable at times but are more acceptable. Whether they are followed by a full point is a matter for house style (see Chapter 7).

The most unattractive abbreviations are those invented by headline writers themselves for use in a tight spot and which, in an extreme case, might produce a headline like this:

'TEC AND 'GROOM
FLEE 'QUAKE
IN 'COPTER

It would be a blessing if all papers adopted the practice of the enlightened few which ban such monstrosities.

Numbers

Multiple noughts look ugly in headlines, apart from being eye-boggling to some, and where the figures are simple the headline can say 'ten thousand' or 'three thousand' or 'two million'. With sterling amounts it is better to say £10,000, and £3,000, although the abbreviation 'm' can be used for million as in £2m. This is often kept in lower case, but looks better in capitals in a caps headline.

Odd figures are usually rounded down in headlines for the sake of tidiness, 5943, for instance, becoming 5900 and 507 becoming 500.

Figures are narrower than their equivalent size letters in a type range, and in Old Style type the 3, 5 and 7 hang below the x-base of the type, a point to watch in line spacing when used with capitals.

CONTENT

Anything that is news to the reader is material for a headline. Yet there are certain points that should be kept in mind.

Taste

It should go without saying that people should not be identified by race, colour or creed unless their race, colour or creed is the subject matter. Words used to describe people should be relevant to the text. If a mother of four is evicted from her house, the size of the family is worth the headline. If a father of four is made editor of a newspaper, the size of his family is incidental.

People can become touchy if certain things about them appear in a headline for no apparent justification. To say that someone appointed to be a magistrate was expelled from school at 15 could be regarded as bad taste, if not actually damaging. If a well-known actor is a homosexual, it is no concern of a reviewer or headline writer. It becomes of concern to the newspaper only if the actor is involved in court proceedings in which his homosexuality is an issue.

Vital facts

It follows from the above that relevance is the important thing in judging a fact for a headline. Not everything that a reporter finds out necessarily concerns the story, but equally the reporter and the subeditor should recognize a fact that will bring a story alive through its relevance.

Take ages. Lists of offenders due to appear before courts usually give everyone's age. It might concern the magistrate that a motorist convicted of dangerous driving is aged 53. It matters little to a newspaper. Yet if a woman of 53 became Wimbledon tennis champion, then her age becomes the headline point. Likewise if a man of 86 becomes a father it is a headline point. Age, like other facts, is important only in its relevance to the story. Effective headline writing lies in the subeditor identifying vital facts such as those that bring a story alive.

The personal touch

The *who* of a headline is important. Opinions and quotations derive their strength from the reader knowing whose they are. There is nothing wrong in having a person's name in a headline, but the abstract *he* or *she* should not be used. If what a person does for a living is the important thing, then give that. For instance: FIREMAN SAVES GIRL WITH KISS OF LIFE, or TOWN'S BAKERS SAYS BREAD IS TOO CHEAP.

Headlines should avoid giving offence by skimping on the name. Mrs Thatcher might accept MAGGIE in a popular tabloid political headline, but diminutives and abbreviated names, such as the American MEG for Princess Margaret are not usually liked. Names like the Pope and the Queen should take the definite article.

Man, girl, baby and mother are acceptable words but the headline should flesh out who or what they are. For example: MOTHER GETS BACK LOVE-TUG BABY; and CONMEN CALLERS ROB A WOMAN OF 82.

'The man who . . .' is a useful way of personalizing a headline – and getting away with a label, as in: MILLIONAIRE WHO ARRIVED LATE FOR HIS HONEYMOON.

In giving someone's opinion in a headline the subeditor should avoid making it seem the newspaper's. If the headline says: ACTION NEEDED TO CURB UNEMPLOYMENT, the opinion must be attributed. Putting the headline in quotes is not enough. Also the news point of an opinion lies in who's it is. If a Conservative Minister says: NEW WAGE DEAL NEEDED FOR THE LOWER PAID, it is a better story than if a Labour Shadow Minister is saying it.

Time

With the *when* of a story, there is a natural assumption in the reader that the news he or she is reading about has just happened or is still happening. The use of the present tense in headlines confirms this, as in TOWN BARBER CREATES HAIR-CUTTING RECORD, and in MANCHESTER SHATTERS LIVERPOOL'S CUP HOPES, even though the game was the day before. BRITAIN MOVES INTO THE BLACK is used in the present tense, despite the fact that the trade figures, to which the headline refers, were for a quarter that ended three weeks previously.

The future can be specified if it is relevant to the story, as in LIZ TAYLOR TO WED EIGHTH, or AT LAST – IT'S A SUNSHINE WEEKEND.

Location

The *where* of a news is most used in an area or regional paper anxious to show the spread of its coverage. Thus a county weekly in Cumbria might have the towns of Whitehaven, Silloth or Aspatria in its headlines to assure these widely spread places that their interests are being looked after. Where an area has its own page of edition news under an area label, the need for place names is less important. In a tightly knit urban evening a few place names are useful where they are relevant to a story but a rash of names on every little filler is tiresome (see examples in Chapters 2 and 3).

In a national daily or Sunday paper, place names are not common in headlines unless their relevance is inescapable as in: WEST RIDING POLICE CHIEF IN NEW ROW OVER PICKETS, or PLAN FOR THIRD LONDON AIRPORT SHELVED.

COMPOSING A HEADLINE

It is possible in composing a headline to reduce the approaches to two, depending on content:

- The direct approach.
- The oblique approach

The direct approach

This entails a factual headline taken directly from the story, and is the commonest, since it conveys the 'hardness' of the main part of the news content. By hardness we refer to the particular quality of stories that is rooted in facts – a Cabinet appointment, a rescue story, a rail accident or a Test match story, for example – rather than stories based on the words, reaction and behaviour of people.

What people say and do can be an integral part of a news story. 'News,' as Harold Evans has said, 'is people.' Yet in the coverage of hard news the quotations from people with whom the reporter has spoken tend to be used in justification, description or corroboration of the facts. It is the facts for which the reporter has gone in search, and it is the facts that he or she has uncovered and checked that form the bedrock of the story and the material for the headline. Thus the hard news of the day might yield such headlines as:

> SIX MINISTERS
> GO IN CABINET
> RE-SHUFFLE

> BOY, TEN
> RESCUES
> PUPPY
> FROM
> WELL

> 17 HURT IN
> RUSH-HOUR
> TRAIN CRASH

> SPEED TWINS
> SHATTER
> ENGLAND'S
> TEST HOPES

The nouns and adjectives – ministers, re-shuffle, boy, ten, puppy, rush-hour, speed, hopes – while falling short of being emotive, are nevertheless evocative of the nature of the happening and are used by the headline writer to make a connection with the reader.

In writing a hard news headline the subeditor is making a qualitative judgement about the story, suiting the words to the story's intended impact.

The oblique approach

Scattered here and there in a newspaper, and sometimes given special typographical prominence, are stories of a softer sort to which we give the term 'human interest'. These are hard to classify but they are generally concerned with the human situation; what someone has done or said, how they have reacted to something; or they concern unusual personal circumstances that have come to light. Often they are about animals and their relationship to humans. There could be an element of oddity or humour

148 *Modern Newspaper Editing and Production*

which lifts a story out of the ordinary, or a quirkiness about the personality of the people involved.

It is here that an oblique approach is useful in writing a headline, though it has to be said that the guidelines are less precise than in the direct approach. A story might inspire a headline based upon a popular song or catchphrase with which the reader is familiar – or a proverb or saying. Irony or humour might shape the subeditor's thinking. The headline could be a quotation perhaps projected in a deliberately wordy way.

Let us look at some examples that have appeared.

JUST THREE
WHEELS ON
HER WAGON
AND STILL
SHE'S ROLLING
ALONG. . . .

This, from the *Daily Mail* is a bizarre story in which a motorist is seen driving along on three wheels after being involved in a crash. It both utilizes wordiness and derives its words from a song.

HAMPSHIRE ARE
FRUSTRATED BY
DILLEY DALLYING

This headline in *The Times* shows punning at work in a cricket report headline about a cricketer called Dilley.

DIANA
RULES
THE
WAVES

in the *News of the World* shows the use of a phrase from the patriotic song in a story about royal fashion.

There is greater scope and more fun for the subeditor writing headlines like this, though most newspapers wisely limit their use. They can be fun to the reader, too, when they are used judiciously and as special 'kickers' to the page, but a glut of them can make news pages seem trivial.

HEADLINE THOUGHTS

To be effective a headline should express one main thought and not hedge it round with qualifications and explanations. In arriving at this thought the subeditor should be looking for what it is about the story that makes it different from any other – why this story is chosen for this particular space on this particular page. If there are other thoughts or facts to be brought forward, they can be given in a second deck of headline – if there is one – or be put into the intro.

If there are two thoughts that are simple and complementary, however, they can sometimes be compounded into one headline, as in PASSENGERS FLUNG OUT AS TRAIN HITS BUFFERS or VIOLENCE FLARES IN BEIRUT AS PEACE FORCE PULLS OUT.

The *after* headline is in a similar category to the *as* headline: CASHIER FOUND GAGGED AFTER RAIDERS SNATCH £50,000, and LOOTERS STRIP SHOPS AFTER FUNERAL ENDS IN RIOT. Inevitably, such headlines are only possible with a bigger character count (see Chapter 9).

Label headlines

These are headlines without a verb, or in which the verb is incidental, and they are not recommended on news. They can sometimes work with colour or human interest stories, as in BRRRRRR . . . GOODBYE, AUGUST! or in THE MAN WITH THE IRON JAW. Possessives can help them to work, as in MANCHESTER'S DAY OF SHAME. A dramatic announcement such as A BOY FOR DIANA stands well because it is really an exclamation.

Some apparently label headlines such as PUMA'S TRAIL IN SNOW gain strength from an implied verb as in this case 'is found' before 'snow'. Generally speaking, label headlines have not the pace needed for hard news and are best left to the features pages.

Split headlines

The old strictures about grammatical 'splits' in headlines – turning lines so that the subject is separated from the object, the adjective from the noun, the adverb from the verb, or the parts of the verb divided – no longer have the force of taboo they used to have. Good flowing headlines were often spoiled by applying these pedantic rules, which date back to the days of wordy, multi-deck headlines in small type.

It is still a good thing not to have the indefinite article, 'a' or 'an', at the end of a line, but only because it is weak visually, not because it breaks a sacrosanct rule. The general scan of the word and the visual effect of the type should be the guiding principles. Does it look right? Does it sound right?

Turn heads

Where a story from Page One (or elsewhere) continues on an inside page, the inside page headline, or turn head, should contain a key word or phrase from the main headline so that it can be identified immediately. This does not mean that it has to be a label. If the story is, for example, about a doctor charged with an offence against a patient, the Page One headline might say: DOCTOR ACCUSED OF RAPING DRUGGED PATIENT. Then the turn head could say: ACCUSED DOCTOR: NURSE TELLS OF 'STRANGE SUBSTANCE IN BOTTLE'. Thus, while the story is clearly flagged for the reader, the headline draws from the inside page material to carry the story forward. Another way would be to use a strap-line:

Accused doctor: hearing continues
NURSE TELLS OF 'STRANGE SUBSTANCE IN BOTTLE'

Either way, the headline should have close to it a line in small capitals, either reversed as a white-on-black or preceded perhaps by an eye-catching black blob, saying

- FROM PAGE ONE

which leaves no doubt that it is the right story.

THINGS TO AVOID

- Avoid headlines with *no, nothing* or *no one* in them. They are crying out to be passed over by the reader. If the story is worth using it must have a positive angle. For example:

    | NO ONE INJURED IN JET TAKE-OFF CRASH | would attract more readers if it said: | 90 ESCAPE INJURY AS JET HITS TREE |

    The positive angle of the escape in the unusual circumstances of a jet hitting a tree becomes the story. The fact that it was on take-off can be given to the reader in the intro. But beware of calling it a miracle escape. A miracle, says the *Shorter Oxford Dictionary*, is 'a marvellous event exceeding the known powers of nature'. The reason ninety people escaped is because the aircraft had brakes and a captain who had the knowledge and skill to use them, thus minimizing the impact resulting from an aborted take-off. This the story can explain.

- Beware the temptation of creating miracle babes, miracle mums and miracle dads in circumstances in which the text offers perfectly rational explanations for what has happened.
- To get the best out of a headline word it should not be repeated in the same headline, nor, if possible, in any other headline on the same page.
- The words *may*, *might* and *could* should be avoided in headlines as they leave the reader uncertain about what happened in the end, unless they keep a look-out for the continuation of the story in the next few days and weeks. This is too much to ask.

If a story is based upon the possibility of something happening it is only worth using if the likelihood is very strong, otherwise the possibilities that can be written about are infinite. Often, newspapers have inside knowledge that something is about to happen which enables them to go a step further than *may*, *might* or *could*. Headlines such as BUILDING SOCIETIES SET TO CUT MORTGAGE RATES or 2 p.c. CUT IN LENDING RATE LIKELY, are ways of writing headlines in which the possibility is almost a certainty. Even then, such headlines should be given to the readers only when the expected event is a matter of days or hours away.

9 HEADLINES: THE TYPOGRAPHICAL CHALLENGE

Modern page design with its 'strength below the fold' allows for a great variety of headline shapes, especially in broadsheet newspapers. The old multi-deck single- and double-column headlines marching across the page, containing as many as thirty-six words in three or four consecutive headline thoughts, have long gone from British papers, lingering on in America in the *Wall Street Journal* and a few conservative country papers. The elaborate rules by which such headlines were constructed, with the main news point in the top deck, are no longer relevant.

Newspapers today commonly go for one-deck, one-thought headlines of from one to about six lines of type depending upon the measure, or width. This, in turn, depends upon their shape and position on the page.

SHAPE

Single-column headlines are usually of two, three or four lines, though sometimes they have as many as five or six if they have dominant end-column positions, or contain some particularly wordy phrase which the page executive wants to put across.

Double-column headlines are usually of two or three lines, generally of a bigger typesize than single-column heads, and are often to be found on the half lead or the third story in importance on the page. The page lead might have a headline that crosses the entire page – a *banner* or *streamer* – running into perhaps a shorter second line alongside a picture or another story.

In addition to these conventional shapes there are long single-line headlines extending perhaps from three to as many as six columns, though not necessarily in large type. These cover stories carried across the pages in short legs in the horizontal layouts now favoured. They are useful when introducing a new item under a wide top-of-the page picture. See examples in Chapters 2 and 3.

While multi-decking is outmoded, the more important stories, both in tabloid and broadsheet layouts, sometimes introduce a second headline thought, either for emphasis or for design purposes, by means of a *strap-line*, or overline as in:

> Sects break truce in Lebanon flare-up
> TEN DIE IN BEIRUT CAR BOMBING

The strap-line can help justify a main headline that seems a little bare on its own, or make possible one that otherwise might not stand at all. The second deck is also back in favour for this purpose on some papers, especially on longer stories that demand headline display, as in:

> STAR'S SAD SECRET
> Soccer ace played on
> although he knew
> wife was slowly dying

Here the method permits a bold main heading to hold the top of the page while justifying it in a wordier second deck.

Another addition to the one-thought headline is the *tag-line*, in smaller type, used to justify an opinion head as in:

> EXPORT BOOM
> 'A TRIBUTE TO
> NEW LOOK
> INDUSTRIES'
> – Chancellor tells the City

More usual than all of these, however, is the headline that concentrates on one point in a single deck. In single-column headlines of this sort the character count in the bigger typesizes might be as little as $6\frac{1}{2}$ letters to the line, although where 18 pt or 24 pt headline type is used down page the count is easier. With wider headlines across two or more columns it is not necessarily greater, for the typesize used is bigger. A seven-column streamer can offer the most difficult count of all. Its one advantage is that the words do not need to break into a number of even lines to satisfy the shape of a narrow four or five-line single-column heading.

A headline that offers a really generous word and character count is the *free-style* one, built usually around a quotation, or statement, chosen because of its humour or apt summing-up of the story. Such headlines are made a focal point in page design, often in conjunction with a picture, and cannot be arbitrarily imposed upon a page without breaching its type balance. For instance, here is a headline on a battle report, which accompanied an atmospheric picture:

> Weapons at the ready,
> faces daubed, no. 2
> Platoon move off into
> the dawn light. . . .

The aim of this sort of headline is to evoke a picture, to give an impressionistic treatment to a news story. It succeeds by being different from other headlines. It signals to the reader that here is a special sort of story.

To have visual balance, type character, and what newspaper designer Alan Hutt calls 'eye comfort', a page of news items cannot be a hotch-potch of headline shapes altered at will to suit whatever ideas a subeditor might have. Fitting the headline thought to a nominated shape and type is a discipline that has to be learnt. Most new subeditors find that the thought and effort involved focuses the mind wonderfully on the facts.

ARRANGEMENT

Within the various shapes of headline, the type is arranged in two main ways. Either it is *centred* on the space available, or it is *set left*, sometimes indented by an en or an em. The reason for choice is readability as much as visual style. A centred headline with a wide difference in width of lines has the eye jumping about to find the starting point of each line and is therefore hard to read. A successful centred headline should have fairly full lines with not too much variation in width, and with the top line full to lend it strength.

A set-left headline is supposed to drop the eye to each successive line more easily. Yet it must still have a full top-line to look visually right, and if the lines are too uneven the raggedness on the right is noticeable on the page. It can give the outside edge of the page a weak look if used in the right-hand end column and is banned from this position in some papers.

A set-right headline is seldom used for the reason that the raggedness on the left makes it hard for the eye to pick up each line as it reads. Look at these examples:

TEARS AS TRAFFIC	TEARS AS TRAFFIC	TEARS AS TRAFFIC
WARDEN TELLS	WARDEN TELLS	WARDEN TELLS
MOURNERS TO	MOURNERS TO	MOURNERS TO
MOVE HEARSE	MOVE HEARSE	MOVE HEARSE

The only other arrangement is the *stepped* headline, still favoured by some US newspapers as in:

> METAL FATIGUE
> IS LIKELY CAUSE
> OF JUMBO CRASH

The first line should be set full left, the last one full right, and other placed evenly in steps between. Only when all the lines are of the same width can this pattern work, otherwise the headline looks off balance. This style is now little used.

TYPE STYLE

In a well planned newspaper the types used on the news pages conform to a regular stock range, with a variant range introduced to give emphasis to a particular story or to form a display focal point, perhaps in conjunction with pictures. As we have seen in Chapter 3, newspapers are careful to stick to their main stock types. If the seriffed Bodoni, or Century, is used for the news pages, perhaps a Gothic or Tempo Sans face will be used as the variant, with a greater, though still controlled, type variety on the features pages. If Tempo or Gothic Sans is the stock news face, then sometimes a little Century Light or the heavily seriffed Rockwell or Pabst (see pages reproduced in Chapters 2 and 3) are introduced as a variant.

Within most type ranges there are bold, light and italic versions, and condensed or expanded ones, etc. There is thus ample variety in size and shape within a type range without any need to mix the ranges.

What is helpful to the headline writer is the modern tendency for newspapers to use only the lower case types of a range, in different sizes and weights. This might be accompanied by the occasional use of capitals in a variant type (Spartan and Century Light are popular for this) for logos or special headlines or second decks. We have already referred to the theory that lower case headlines, with their flowing outline and more varied contour are easier to read at sight, as in ordinary reading text. Along with the tendency towards lower case is a decline in the use of italic type, both in lower case and capitals, for the same reason – readability.

Using lower case types certainly eases the letter count and is helping to free headlines from dependence on short stereotyped headline words of the sort much used in tabloid papers, with their narrow columns and large caps type. The other side of the coin is that bigger letter counts can result in headlines which are wordy and lack punch and eye appeal unless the advantage is properly used. And no advocacy of all lower case headlines can dispute the fact that the success of tabloid layout was built on the eye-catching poster effect of big sans caps headlines.

Character counts

To get an exact count, even though familiar with the type, the subeditor consults the office typebook, of which all production journalists are given a copy. Each newspaper or print shop, whatever typesetting system they use, publishes one of these to display the type ranges available. The examples show each size in capital and lower case letters, with roman and italic, expanded and condensed versions where applicable, along with numbers, fractions, punctuations, stars, blobs, and any other characters in the range.

Specimen typesizes are shown in column widths and multiples of columns to match the format of the newspaper, so that counts can be made against widths of single-column, double-column, three-column and so on. It will also be seen that letters within a size vary in width, not only in their lower case form but in their capital letters, too. The count for a headline has to take account of this variation and also of the letter space needed between the words.

A headline count is easier using capitals. Apart from the I which counts as a half character, or the M and W which each count as two, the variation in width among the remainder is slight enough to be self-cancelling between one letter and another. Allowing the space between words as one character, it is easy enough to get an accurate count in most types (though some do have odd-width letters). Even so, a rash of Is or Ws in a headline can lead to imbalance, however carefully the characters are counted.

In the lower case letters there is more variety of width, with f, i, j, l, r and t all being narrower than the average width, and averaging out at about half a character each, while w and m count usually as one and a half. Punctuation can vary from a half character to a full character, as between a comma and a question mark. Dashes and ellipses invariably take two or more character spaces.

Most photosetting systems are formatted to allow single character spacing between words, although this can be reduced to a half letter to accommodate a much wanted headline, or otherwise save space. The readability with reduced space depends on whether the letters separated by the space have upright or receding strokes, the latter giving the impression of greater white, as with the letter c or with capitals, A, V and W.

Working out a headline count goes like this:

> ISRAEL IGNORES
> $\frac{1}{2}1111111\frac{1}{2}111111 = 13$
> UN ARMS WARNING
> $111112112111\frac{1}{2}11 = 16\frac{1}{2}$

The second line here is too long for a headline which allows a maximum count of 14 so 'UN' is deleted, leaving the headline as:

> ISRAEL IGNORES
> $\frac{1}{2}1111111\frac{1}{2}111111 = 13$
> ARMS WARNING
> $112112111\frac{1}{2}11 = 13\frac{1}{2}$

This gives two almost equal lines with a little white each side when centred on the space. But watch out when numerals are used. They are narrower than letters. The odd one or two will make little difference as in

>   TEACHER
>   CHOOSES
>   PRISON
>   INSTEAD
>   OF 'UNFAIR'
>   £10 FINE

But in the following, the last line will be shorter than expected:

>   SHOP-LIFTER
>   TOLD: 'PAY
>   BACK FIRM'S
>   £50,000'

The traditional method of type usage based upon the standard sequence of sizes of 7 pt to 14 pt, for text, and 14 pt to 120 pt for headlines is not essential in photosetting systems. Since the type is produced as a photographic image it can be reduced or enlarged to any size within the top and bottom limits of the system. Yet the standard sizing common to the old hot-metal systems is still universally used and remains a useful yardstick in visualizing. Systems are formatted to produce type in standard sequence. This enables pages to be given a calculable type balance and controlled eye appeal at the planning stage. The subeditor also knows the sort of letter count that is wanted and has a visual picture of how the headline will look.

A useful thing from the headline writer's point of view, when editing electronically, is that there is usually a 'headline count' key which can be pressed to show on the screen whether the headline fits, or is too strong and by how many characters. Thus time is not wasted by consigning a doubtful headline to the photosetter.

Using a typebook gives a subeditor the 'feel' of type, and helps in the understanding of projection and page design. A production staff with a feeling for type and an awareness of its possibilities is the most important requirement for a good-looking newspaper. A newspaper's type character is threatened if subeditors are blindly writing headlines they cannot visualize or connect with the paper. In this respect the method used by some newspapers of referring to frequently used headline sizes and shapes by graded numbers or letters can be counter-productive. Though quick and convenient for the setting system it is discouraging to the young journalist keen to learn about typography and production.

The typebook is the young journalist's introduction to the world of typography. The usage and nuances of effect of different type ranges and

PLACARD								
		× 1	× 2	× 3	× 4	× 5	× 6	× 7
144	CAPS	1½	3	5	6½	8¼	10	11¼
120	CAPS	1¾	3¾	6	8	10	12¼	14½
	l.c.	2	4	6	8	10	12	14
96	CAPS	2	4½	7½	9¾	12½	15	17¼
	l.c.	2¼	4¾	7	9½	11¾	14	16½
72	CAPS	3	6¼	9¼	12¾	16	19	22¼
	l.c.	3¼	6½	9½	13	16	19½	22½
48	CAPS	4½	9	13½	18	22½	27¼	31½
	l.c.	5	10	15	21	26	31	36

Figure 28 *Type charts giving tabulated counts can be compiled from the office type book for quick reference in headline writing*

variations of letter shape can convey to the interested a feeling for the 'psychology' of type; an awareness that there are fast and slow types, and male and female, that certain types are suitable for certain functions.

While the typebook remains important, casting off headlines can be speeded up by the use of a type chart giving counts and measures in tabulated form. Figure 28 shows part of a type chart which gives counts in capitals and lower case letters for Placard, a condensed heavy sans type used in some tabloid news headlines. The counts are shown from one to seven columns of a standard tabloid page seven-column format.

Another useful idea for young subeditors anxious to get the feel of type is to cut out examples of the more frequently used headline shapes in the paper and stick them on a card with notes of the count and the type name and size.

Once involved in page design a journalist with a flair for visualizing can learn how to give weight and boldness to a news page, authority to a leader page, and busyness to the sports section; how a condensed type used sparingly can enliven a standard format layout, and yet how a newspaper's type character is maintained by consistency.

SPACING

Tight character counts pose problems not only with the words in headlines but with the visual balance of the lines. White space used sparingly and creatively in a layout can help the eye, but white space in a headline that was intended to be occupied by type is wasted. It does nothing for the page and, if occurring frequently, can make it look empty.

A useful rule with headlines is that the letters should, as far as possible, fill the space. If white is wanted for visual reasons the lines can be indented (i.e. set narrower than the space to be filled) or a deliberate display can be projected with planned white space as part of the pattern. In headline writing, the use of white, it should be remembered, is to give adequate separation between words and lines so that the headline is easy to read and the visual effect satisfying.

For instance, a headline should not finish up looking like this:

BOY
TEN, RESCUES
DOG FROM
WELL

Nor should it become pyramid-shaped:

SPEED TWINS
SHATTER ENGLAND'S
HOPES AT TRENT BRIDGE

or be a staircase:

RAIN MAKES
AUGUST BRITAIN'S
WORST FOR 50 YEARS

If achieving a good shape is a problem then the best thing is to change the words and start again.

The bigger the type the greater the damage caused by bad spacing. An 84 pt banner that falls short by half a column at each end weakens the top of the page and loses authority. Likewise, long single line headlines below the fold need to be full if they are to dominate as they are intended to.

Vertical or line spacing is automatic with most typesetting systems but it helps with headlines if space can be given visually in lower case lines where ascenders and descenders clash one on to the other. Lower case type is difficult to space tidily. Spacing lines of capitals is a matter of consistency, but beware the intrusion of the cross-line descender on the capital Q.

Spacing through the lines of a headline just to fill up space at the page make-up stage can result in poor visual balance and should be avoided except when time is short. It is better to extend the text.

Apart from the above general strictures, spacing is a matter for visual style. There can be no universal rules in line spacing of headlines or columnar spacing. Some magazines make a feature of fairly tightly spaced lines which are easily achieved in photoset systems and can look good in gravure or in advertising display. Others go for air or massive front indents. Some newspapers indent all columns by a nut each side to give more vertical separation of matter, or perhaps indent the front of the lines only, as setting style. Others print the text with type that has a bigger body – for example 7 pt on 8, or 8 pt on 10 – thus marginally increasing the separation between the lines of the text.

The golden rule is to be consistent. A clean page is better achieved if the line spacing of headlines is not tight here and wide there. There is little point in departing from normal columnar and text spacing, whatever the width of the measure, except in cases of special display such as in cut-offs and panels. Maximum readability demands that one line should be clearly differentiated from another and one word from another, but that otherwise the eye should not be aware of the white space.

Other points to watch:
- Quotes: These can look effective in set-left headlines if left hanging in the white rather than being counted in with the characters, where they unbalance the effect at the beginning of the line.
- Underscoring: This should be used sparingly to draw attention to a line or lines. In lower case type the underscore should be spaced evenly from the bottom of the x-line and not break the descenders.

THE CREATIVE MOMENT

There are two methods of thinking your way into writing a headline. One is to read the copy and try to come up with the wording before subediting the story. This is necessary if the headline is to be the keystone of the layout. The method has the advantage that the headline is written at a stage when the type and even the layout can be adapted to accommodate it.

The other method is to subedit the story, absorb the material in the process and then work out the headline. The disadvantage here is that by now you are closer to the copy deadline and time might be short.

Some stories produce a headline easily so that in reading the copy it jumps into your head, while others are difficult to transmute in a few words. Some human interest or 'mood' stories can be like this when the hard facts are few and the effect is in the telling or the description, as with royal visits or elaborate ceremonials. In such cases, and especially when the letter count is limited, it helps to jot down likely words on a copy pad and shuffle them, even with screen subediting. Sometimes a half dozen key words emerge and the job is to arrange them in the right order to give balance and sense, as in a puzzle.

Take an imaginary situation. A teenager called Janice Thompson has gone missing from home. Her friends tell the school head that the girl had quarrelled with her mother over drugs and glue-sniffing. A reporter on the local paper, which has already been told by the police that the girl is missing, hears about this and goes to see the mother. She is anxious to talk. She says that Janice, fifteen, walked out after being told she was no longer welcome in the house if she continued to smoke pot and sniff glue. The reporter files the story and it is given a strong top-of-the-column space on Page Five.

The story is late. The subeditor checks it, casts it off quickly and writes down the words: *mother, teenager, drug, quarrel, vanish* and *home*. The first effort is:

TEENAGER
VANISHES
AFTER ROW
AT HOME
OVER POT

This is no use, for the fact that the girl has gone missing is not new. Also the word 'after' splits the thought into two.

MOTHER'S BAN
ON POT DRIVES
TEENAGER FROM HOME

is better. Nearly there, in fact. How does it look as five lines of single-column?

MOTHER'S
BAN ON POT
DRIVES
TEENAGER
FROM HOME

Still a bit slow. Also the second line and last lines are one character too long for the typesize. Using the name would connect the story with the readers. Try again.

MISSING
JANICE:
MY DRUGS
BAN TO
BLAME
SAYS
MOTHER

Too long, but better, although the headline thought has again been split. Try not to use the colon. And why not 'mum' instead of 'mother'

> MY DRUGS
> BAN DROVE
> JANICE
> FROM HOME
> SAYS MUM

in the five lines and just right. Fill out the middle line by making it JANICE, 15, and we're there.

In another story a foreman at the local car factory is sacked after an altercation in which he hits an apprentice and knocks him unconscious. The apprentice had accused him of being a 'scab' (a non-striker) during a strike.

The words on the pad are: *union, sack, foreman, angry, scab, factory*. First of all, the 'scab' jibe is not new, or contested. The new issue is what is to happen to the foreman. The first effort at a double-column two-line headline, therefore, is:

> FOREMAN LOSES JOB
> IN 'SCAB' JIBE ROW

'Scab' jibe row is not liked. Also the headline needs more pace. A neat shuffle produces the answer:

> 'SCAB' JIBE COSTS
> FOREMAN HIS JOB

There is nothing wrong with writing a headline in this order. Having subbed the story first, the subeditor knows thoroughly what it is about so that whatever headline emerges has to square with the story, as opposed to writing a headline first – which might be based on a superficial knowledge of the story or even, in some cases, on a misconception acquired at first reading. There is a subconscious tendency, too, where a good headline has cropped up first, to bend the story towards the headline. This can lead to distortion.

ALTERNATIVE WORDS

'If there is one outstanding fault in present-day headlines it is that very sameness with which they speak day in and day out. Much of this monotony is inevitable because of the cramping space limitations that compel repeated use of words, phrases and constructions that have been found most serviceable.'

Garst and Bernstein, *Headlines and Deadlines*.

Headlines: the typographical challenge 163

It will be seen from the examples quoted in this chapter that the limitations on word length and the search for type balance lead the subeditor to use a short word where alternatives are available. Fortunately, this offers few problems. There is almost always a shorter word. The danger arises when certain useful short words are allowed to become clichés: words to fall back on as familiar 'get-outs' even when the facts of a story do not entirely justify their use.

If is a favourite occupation of critics of newspaper headlines to tot up how many times in a day's issue of the national tabloid papers verbs such as axe, snub, lash, probe, rap, slash, hit, quiz and quit can be found.

Using the same truncated words can lead to tired don't-read-me headlines, which is a pity with a language as flexible as English. Young subeditors will find that the best way out of this trap is to devise their own list of short alternative words for use in headlines. Roget's Thesaurus of English Words and Phrases, a good modern usage dictionary such as the Penguin English Dictionary, and a stout exercise book are all that are needed for this assignment.

Here, to ease the way, are some suggested alternative words, but remember in using synonyms that there can be nuances of meaning between one and another, and that a story could be misrepresented if an alternative word were used without proper thought. Also, as the list shows, some words have two or more meanings, each with its synonyms.

A
abandon
desert
drop
give up
leave
neglect
pull out
quit

abbreviate
cut
lop
reduce
shorten

abolish
close
drop
end
finish

rid
squash

abscond
flee
leave
run

accelerate
hasten
hustle
press
push
quicken
rush
speed

accumulate
amass
build up
gather

hoard
store

acquire
buy
collect
gain
get
grab
take

adjust
alter
change
revise
shift

administer
boss
control
direct

give
manage
rule
run

aggravate
annoy
inflame
irritate
provoke
worsen

agreement
accord
bargain
bond
deal
pact
treaty

alleviate
ease

lessen
let up
lighten
reduce
relieve
soften

allocation
lot
part
portion
quota
ration
share

amalgamate
combine
fuse
join
link
merge
mix
team up
unite

amalgamation
joining
link
merger
tie-up

announce
disclose
notify
proclaim
report
reveal
state
tell

appeal
ask
call
plead

appoint
choose
invest
name
pick

appointment
job
mission
place
post

apportion
allot
distribute
divide
share
split

appreciate
enjoy
grow
increase
like
rise

appropriate
grab
loot
nick
seize
snatch

approve
agree
allow
OK
pass
permit

arbitrator
go-between
judge
mediator
referee
umpire

argument
dispute
quarrel
row
rumpus
set to

ascertain
check
confirm
discern
ensure
find out
inquire
learn
seek

assistance
aid
back-up
help
relief

attain
achieve
get
reach
secure

authorize
allow
approve
back
favour
let
permit
sanction
sponsor

B
beginning
birth
dawn
debut
onset

opening
start

bequeath
allot
give
leave
provide
will

bewilderment
daze
puzzle
shock
surprise

business
company
firm

C
calculate
assess
estimate
rate
tot up
value

challenge
contest
dare
defy
dispute

change
alter
amend
revise
reword
shift

circumvent
balk
foil
outwit
skirt

close
call off
end
finish
shut

commence
begin
embark
open
start
take off

communicate
call
pass on
reveal
tell
write

competition
contest
fight
game
rivalry

complain
bitch
groan
grouse
grumble
moan
nag
object
protest
resent

confront
challenge
face
halt
present

congratulate
commend
praise

consider
examine
inspect
look at
mull
probe
report on
study

construct
assemble
build
erect
form
make
put up
raise

continue
extend
go on
persist
plod on

contradict
deny
dispute
dissent
gainsay
oppose
refute
reject

control
curb
limit
organize
peg
run

cooperate
help
join with
share

create
build

design
form
invent
make

criticize
assess
censure
challenge
chide
decry
evaluate
rap
rebuke
review
slam

D
damage
harm
hit
hurt
spoil
sully

deflate
contract
pinch
shrink
squeeze

demonstration
demo
march
show
sitdown
sit-in

deprecate
belittle
damn
discount
discredit
knock
run down

description
account
picture
sketch
story
tale

designate
appoint
fix
name
select

destroy
break
end
ruin
shatter
smash

disagreement
clash
conflict
dispute
quarrel
rift
row
rumpus
wrangle

discriminatory
biased
partial

distribute
circulate
give out
issue
supply

division
break-up
parting
split

donation
gift
grant
hand-out
present

E
employment
career
job
post
task
work

endorsement
acclaim
approval
backing
sanction
support

essential
key
main
necessary
vital

exaggerate
blow up
boost
distort
magnify
overstate

examination
analysis
inquiry
probe
scrutiny
study
test

exonerate
absolve
acquit
clear

expedite
hasten
hurry
push
quicken
rush
speed

explanation
account
alibi
answer
excuse

F
fabrication
falsehood
invention
lie
story
tale
untruth

facilitate
advance
ease
expedite
smooth

forbid
ban
bar
block
deny
disallow
prevent
reject
stop

foundation
base
basis
charity
grounds
origin

roots
trust

fundamental
basic
primary
vital

G
govern
command
control
direct
manage
rule
run

grievance
grouse
grudge
injury
wrong

guarantee
(*noun*)
bond
pledge
promise
support
surety
warranty
(*verb*)
endorse
ensure
pledge
warrant

guillotine
axe
chop
cut off
gag
silence

H
hallucination
delusion
dream
freak-out
illusion
mirage
vision

hazardous
dangerous
dodgy
perilous
risky
unsafe

I
illegitimate
banned
illegal
illicit
lawless
unlawful

illustrate
explain
picture
show

imminent
at hand
near
soon

important
big
great
high
key
notable
prime
top
vital

improve
better

enhance
heighten
polish
restore

improvement
advance
betterment
progress
recovery

inaugurate
begin
install
launch
open
set off
start

independent
free
impartial
neutral
unbiased

influence
colour
induce
lead
push
sway

inquire
ask
beg
look into
question
seek

instigate
cause
incite
provoke
set off
spur
start

interfere
hinder
interrupt
meddle
pry
thwart

interrogate
ask
examine
grill
pump
question
quiz
vet

introduce
explain
float
launch
promote
start

investigate
analyse
check
delve
examine
inquire
probe
pry
sift
study

invite
ask
beg
bid
call

J
jeopardize
endanger
hazard
imperil

risk

judgement
decision
finding
result
ruling
verdict

justify
back
bear out
confirm
defend
endorse
excuse
explain
support

K
kidnap
abduct
capture
carry off
grab
seize
snatch
steal
take

L
leader
boss
chief
head
master
ruler
guv'nor

legalize
allow
ordain
permit
ratify

luxurious
costly
lush
plush
posh
rich

M
maintain
assert
hold
keep
insist
support

management
board
company
directors
executive
firm
owners

manoeuvres
dodges
exercises
plots
ruses
tricks
wiles

manufacture
build
construct
make
produce

mediate
decide
intervene
judge

moderate
allay
control

lessen
limit

modification
alteration
amendment
change
switch

N
negotiate
bargain
confer
discuss
haggle
parley
talk

neighbourhood
area
district
environs
locality
place
section
zone

nominate
appoint
call
choose
decree
invest
name
present
propose
put up
raise

O
objective
aim
ambition
end
target

observe
check
eye
inspect
look at
note
see
spy
watch

operate
act
control
run
work

oppose
bar
block
challenge
combat
contest
counter
fight
obstruct
rebuff
reject
repel

organize
control
develop
fix
form
plan
run
set up

P
pacify
allay
calm
cool
ease
heal

quieten
settle

permanent
abiding
constant
durable
enduring
lasting

postpone
delay
hold up
put off

prohibit
ban
bar
disallow
end
forbid
halt
prevent
rule out
stop
veto

prohibition
axe
ban
embargo
end
halt
veto

proposition
idea
notion
offer
plan

purpose
aim
ambition
end
idea

plan
reason

pursue
chase
dog
follow
harry
hound
hunt
search
seek
trace
track
trail

pursuit
chase
hunt
quest
search

Q
question
challenge
doubt
point
probe
query

R
recommend
back
boost
commend
praise
present
support
urge

reduce
axe
chop

cut
lower
slash

regulate
control
govern
legalize
order
run
vet

relinquish
abandon
drop
forego
give up
leave
quit
resign

repudiate
deny
disclaim
spurn

requisition
acquire
get
grab
seize
take

resign
abandon
give up
lay down
leave
quit

responsibility
duty
job
role
task

S
sanction
allow
approve
OK
pass
permit

scrutinize
check
examine
inspect
study
vet

settlement
bargain
bond
deal
pact
treaty
truth

support
aid
back
foster
help
prop
push
stand by
uphold

T
terrorize
awe
bully
frighten
scare
threaten

tolerate
accept
allow
endure
live with
stand

U
unblemished
clean
faultless
flawless
guiltless
spotless
virgin

unyielding
firm
fixed
resolute
solid
steady
stout
unmoved
unshaken

undermine
belittle
impair
sap
weaken

undertaking
agreement
bargain
deal
duty
mission
plan
promise
task

V
vacillate
demur
hedge
hesitate
sway
waver
wobble

vanquish
beat
defeat
rout
scatter
shatter
smash
trounce

vindicate
bear out
justify
set right
uphold

vulnerable
delicate
exposed
suspect
tender
weak

10 SUBEDITING: FURTHER TECHNIQUES

In editing copy the subeditor must ensure not only that a story is accurate and is clear and readable, and of the right length, but that it is legally safe to print. Production delays caused by a late decision to take a story out of a page because it might be libellous can cost money and lose sales.

All newspaper companies retain a trained lawyer or a senior journalist versed in the law, to read copy and advise the editor of potential legal traps. If a story is considered unsafe to use, the lawyer or legal reader, after consulting the editor or chief subeditor, will issue a *legal kill*, which means that the story must be thrown away, or removed from the computer.

Any legal alterations to make a story safe are passed to the subeditor who marks them 'set catch *legal must*' when re-entering the corrections into the system for typesetting. This catch-line, which is removed at make-up stage, is to remind the make-up person that the corrected part must not be cut or left out.

Electronic editing causes no particular problem to the legal reader. While security of text means that computer access is limited to particular areas for those performing editorial tasks, the legal reader has access to all incoming texts, which can be called up, file by file, and read on screen. Even so, it is necessary for the legal overview to extend to the finished page and it is usual for photocopies, or 'proofs' of each one to be checked before being 'sent' to ensure that no legal danger has resulted from editing, captioning or headline writing.

The over-view by the lawyer, or legal reader, is a general one and varies in efficiency from office to office and it does not absolve the subeditor from watching carefully for legal dangers in copy. If there are any doubts about a story that cannot be resolved in the editing the subeditor should refer it in good time to the legal person before consigning the copy to typesetting, thus saving time. A decision can then be taken as to whether there is a danger or not and, if need be, the story can be made legally safe.

The difference between a safe story and action for damages can often turn on a phrase or the arrangement of the words at a critical point in the story.

Lessons in law for newspapers are included in all journalism training schemes, but it takes time for a young subeditor to develop the sixth sense an experienced colleague uses to spot legal traps that can lie behind innocent-seeming words. Re-writes of stories are an obvious danger area, when facts are being rearranged and quotations paraphrased or perhaps written down wrongly. Yet a 'tick-marking job', a well written story that needs little editing or cutting to fit, can hold greater danger because it lulls the subeditor into a state of false confidence. The effect of every word or phrase needs to be properly weighed and a story that seems to need little attention should be read through again carefully.

SUBEDITORS AND THE LAW

Two thoughts should be kept in mind by journalists when considering legal problems in editing: one, that a mistake or damaging statement gains much more currency with the public in a newspaper than by any other means of dissemination and therefore can cause greater harm; two, that newspaper companies are regarded as rich and a good target for litigation. Editors are thus sensitive to those laws of which newspapers are most liable to fall foul.

There are two sorts of laws that affect newspapers in Britain:

1. The general law of the land to which editors are liable in the same way as is the ordinary citizen. These include the laws covering defamation of character, or libel; contempt of court, trespass, confidentiality, and the various provisions of the Official Secrets Act, which restrict the passing on or circulating of certain information.
2. Laws aimed more specifically at the press and broadcasting media to restrict the publicity given in court cases and which are, in effect, a type of censorship. These include the laws forbidding publication of evidence in divorce and other matrimonial cases, the identification of offenders in juvenile cases, the publication of evidence given at lower courts against people committed for trial to higher courts, the details (and names in some cases) of people involved in sex offences, and also some provisions contained in the race laws.

While all these laws worry editors and news editors since they restrict what a newspaper can say and do, not all of them concern the subeditor engaged in the daily production of the paper. The laws restricting the coverage of court cases are accepted and applied. While a subeditor should be aware of types of reporting restrictions, he or she is not likely to be presented with a detailed blow-by-blow account of the evidence in a steamy divorce case. The evidence is not covered because of the legal restrictions, and editors are aware that only the names and judge's summing up in such cases can be used.

Nor are the laws covering juvenile court cases flouted. Names and identification of juveniles can be given only if the magistrates direct that they should be published in the cases of very serious offenders who put the public in danger. Here a magistrate's direction becomes an important part of the reporter's story.

The provisions of the Criminal Justice Acts of 1967, 1980 and 1981, which forbid the publication of evidence against people sent for trial from lower courts (except in special cases), though disliked by editors, are universally applied, as are those of the Sex Offences Acts.

Yet legal traps can still lurk in copy. Chief of these is the perennial danger of libel in which the threat of being sued can cause a timid editor to shelve a story which should have been published for the public good.

Libel

A newspaper is guilty of libel when it can be proved that a person's character or livelihood has been damaged as a result of statements made in the paper. A successful action can result in substantial damages being paid. While some well-known cases get much publicity, many more are settled unknown to the public by out-of-court arrangements in which quite large sums of money change hands, accompanied by a printed apology. This method at least avoids heavy court costs in cases where a newspaper feels the judgement might go against it.

Defence against libel is difficult. An editor might plead the truth of the statements or his or her defence might be that they are not defamatory or that the story was 'fair comment made in good faith and without malice about a matter of public interest'.

While the overall decision on whether or not to publish a story is the editor's, the subeditor has to be careful in the arrangements of words and facts in a story in which a known danger of libel lurks. This is particularly so in stories where the editor, having taken legal advice, is ignoring a letter threatening libel action if the story appears. Such threats can be devices used by desperate people to stop publication and are ignored only if the editor has good legal grounds (and a good legal adviser). In such a story the office lawyer should be allowed to see the edited version at an early stage to vet it in time to avoid production delays.

A persistent litigant or a wily lawyer can find libel in the most innocent statements and no paper is ever free of writs. Some, however, are just attempts to extract money out of the company by unscrupulous people, and a newspaper learns how to deal with these.

In stories of known legal danger, the subeditor has to make sure that risky statements are checked and corroborated; that people against whom accusations are made are given space in the story to make their reply; and that background details for someone in the news are not chosen out of malice to make them look small or ridiculous. Any obvious danger points should be checked back with the reporter.

People with criminal records are particularly sensitive about having them mentioned in stories that have nothing to do with their murky past. Actors can stand criticism of a particular role but could sue if their general professional competence were being questioned. A person in public life would expect to meet opposition and criticism but it would be actionable to say they were not fit to be in public life. Likewise, suggestions that people are drunks or take drugs can be dangerous, or stories that suggest they have been less than honest in their handling of public or shareholders' money.

Contempt of court

This means broadly any conduct or spoken or written words or printing of pictures which might impede the working of a court or bring justice into disrepute, and it is a law aimed at everyone, not just the press. Yet the press, because it publishes the proceedings of courts, is particularly exposed to the danger of being in contempt.

Here are some of the things that a subeditor, and anyone else involved in editorial production, must watch for:

1. No picture should be published of a person accused or expected to be accused of an offence until he or she has been identified in court. An exception would be where the police have issued a picture of a wanted person. Even here the description attached to the picture must avoid accusing the person and must contain words such as 'is wanted by the police for questioning in connection with . . .' etc.
2. A newspaper must not publish new facts or evidence about people being tried before a court while the trial is in progress. The people charged are not in a position to refute them, the defence or prosecution case might be damaged by them and the jury influenced.
3. The newspaper must not try to interview any witness or person involved in a trial. The printing of such an interview can put the paper in contempt.
4. Any criticism of the judge or the court proceedings while a trial is in progress is considered serious contempt.
5. No attempt must be made by a newspaper to get in touch with a member of the jury during a trial.

These are the main points concerning news stories about court proceedings but there is a wide area beyond this in which, at the decision of a judge, a newspaper might find itself in contempt of court (and open to the penalties of imprisonment and fines that go with it). For instance, journalists have gone to prison in Britain for refusing to disclose to a judge the sources of information contained in a story about an accused person printed before a trial began. Also, the proceedings of judicial tribunals, presided over by judges, into disasters and other situations have been deemed to be subject to the laws of contempt of court, thus restricting newspapers in which they can print in the same way as with court proceedings.

The idea of the laws of contempt of court is that the person, once accused, should be able to get a fair trial in front of magistrates or a jury. For example, a persistent pattern of crime in an area, such as attacks on women or children, can produce in the public mind the shadowy figure of a marauder who becomes personalized as the Yorkshire Ripper or the Beast of Bournemouth. As long as the police search goes on and further attacks are reported the name Yorkshire Ripper or Beast of Bournemouth dominates the headlines. Once a person has been arrested and charged, however, or even if the person is being held for questioning and not yet named, the editor is in danger of being in contempt of court by associating the name of the marauder with the person held.

The matter has, in legal terms, become *subjudice*. It is subject to the due processes of the law, which must be allowed to proceed without outside interference. From this point a newspaper story can give only those details that are allowed under the law.

The sort of story that first appears after someone is arrested, might run like this:

A man was helping police with their inquiries at Extown police station last night in connection with the deaths of two teenage girls from Extown whose bodies were found last Wednesday in the River Ex.

The next development of the story might be:

Charles Jinks, a labourer, of Caxton Street, Riverport, was charged in Extown Magistrates' Court this morning with the murder of Elsie Jones, of Privet Street, Extown, on or about October 8.

He was given legal aid and was remanded in custody for further inquiries. Bail was refused.

It would be wrong of the local newspaper to introduce the death of the second girl into the story at this stage, even though they might know that a further charge was pending. It has to wait.

Eventually Charles Jinks is charged in the Magistrates' Court with the murder of both girls and evidence is presented and pleas taken. The paper can still publish only what is allowed under the Criminal Justice Acts, which means that, unless the magistrates direct otherwise, only Jinks's name and address, the charges against him, the names of the girls, his plea and the fact of his committal to a higher court can be given.

Not until the case comes before the Crown Court for the area and the prosecution and defence cases are given in full before a jury can the evidence be published in the press. Only after the trial has finished and judgement been given can the paper comment on the case, interview people involved and – if the evidence shows it to be justified – use once more the term Yorkshire Ripper or the Beast of Bournemouth. It can then, if it wishes, even comment on the way proceedings were brought and on the police handling of the case.

By observing the law of contempt of court, the editor has ensured that the paper's coverage has done nothing to impede a person's fair trial (since under British law a person is innocent until presumed guilty) by publishing information or comment that might have influenced the jury or damaged the prosecution or defence cases.

The job of the subeditor is to check that the law has been observed by the reporter in his or her account of the hearings at the various stages and not to allow anything into the newspaper that will get it into trouble.

The Official Secrets Act

This is a wide act which can be used to cover many aspects of Government business as well as security or state secrets. Controversial areas are spelt out to the press from time to time in D-notices which are requests to the press not to publish certain things on threat of being guilty of breaching the Act. Punishment is not automatic for ignoring D-notices but an editor would have to have good grounds for doing so. The decision here, for editorial production, is whether to publish an item at all, and it is one for the editor.

Newspapers would be in trouble with litigants much more were it not that under British law the need for freedom of speech in the public interest is recognized and certain newspaper reports are protected by 'privilege'. The law on privilege covers the reporting of court hearings and sittings of Parliament and public bodies where serious accusations can be made about people which, if made and reported elsewhere, could lead to actions for libel.

The application of privilege needs to be learnt by journalists during training for it is a valuable guide to the protection their reports have. It must be known by subeditors. Privilege comes in two sorts.

Absolute privilege

Absolute privilege covers 'a fair and accurate report of judicial proceedings published contemporaneously' provided that legal restrictions on coverage in certain cases specified in law have been applied. However untrue in fact statements in court are, or however unfair, a newspaper who reports them and says who made them is safe from action for libel. An exception has been statements involving sedition, blasphemy and obscenity, though these have become increasingly hard to define.

Points to watch:

1 'Fair and accurate' means that the report must be fair to both sides.
2 The report is not covered by absolute privilege if it includes matter gleaned from documents that have not been read out in court.
3 Protection does not include anything not part of the proceedings, such as an outburst or scene in the courtroom.
4 The report must be 'contemporaneous'. This means it should be carried in the next issue of the paper after the hearing.
5 The wording of the headline must not go beyond the story.
6 Beware of wording. 'A 'charge' of murder only becomes a 'case' of murder after the verdict. If a person at a lower court 'elects' to go for trial, do not say he or she was 'sent' for trial, which suggests the court has decided there is a case to answer. A statement made by a person is an 'alleged statement' until it has been agreed and accepted in court.
7 Descriptions of witnesses, their dress or behaviour, are not covered by privilege.

Qualified privilege

This allows protection for reports of judicial proceedings that are non-contemporaneous, and to reports of Parliament and other public bodies over a wide field. Broadly, qualified privilege means that the reports are privileged in law provided there is no malice or other improper motive behind publication. Where these motives can be established in court, then the protection of qualified privilege is of no avail. This is an important distinction that needs to be kept in mind by subeditors dealing with such reports.

The areas covered by qualified privilege include: Parliament, Commonwealth legislatures, international organizations and conferences to which the UK Government sends a representative, public inquiries, bodies formed in the UK to promote the arts, science, religion or learning, or the interests of trade, business or industry; associations for promoting sport and pastimes to which members of the public are invited, public meetings held for a lawful purpose, meetings of local authorities, committees or tribunals appointed by Act of Parliament, and reports of notices or information issued to the public by the Government, local authority or the police.

Schedules listing these and other occasions of qualified privilege, and also spelling out some important exceptions, are published as part of the 1952 Defamation Act, and its successors.

For the subeditor, danger can crop up with statements made by some public bodies covered by qualified privilege in which the background material to which the newspaper might have access, is not so covered. The advice of the office lawyer, or legal person, is the only answer in some cases.

The publication of material not covered by qualified privilege does not mean that the newspaper is in trouble automatically – only that if a court action follows, qualified privilege cannot be used as a defence.

REWRITES

Subeditors are generally thought by reporters to be too keen to rewrite copy. It can be galling for a reporter who has taken care over writing up a difficult assignment to find that half the words in the published version, perhaps under his or her name, are unrecognizable.

It is wrong and unjustifiable for a subeditor to alter and rewrite copy just for the sake of it. On a well run subs' table this does not happen. For one thing, subeditors should be too busy editing to deadlines to devote time unnecessarily to a story. For another, the chief subeditor would not stand for a well turned story which earned its space as written being pulled apart.

Yet rewriting is necessary just as heavy editing can be necessary for reasons that have nothing to do with the way a reporter has written the story. The treatment of copy, as we have seen, is influenced by the volume and changing pattern of news on the day, and the availability of space. It is also governed by new developments that put a story in a different light, the use of other copy sources, a change of news angle to suit a night editor's or chief subeditor's judgement, or simply a clash with editorial policy.

It can happen that the work which needs to be done on a story for these reasons is so complex that rewriting is quicker and simpler than trying to produce a readable sequence from fragmented heavily edited sections.

The following examples show where rewriting would be the best means of editing.

Angling

There is nothing sinister about angling a story. It simply means deciding the viewpoint from which to tell it. The intro, in effect, nominates the angle. The fact sequence that follows supports and justifies it.

It can happen that the night editor or chief subeditor, with more mature judgement, sees a story differently from the reporter, and the subeditor is instructed to 're-nose' it from a new angle. Here are some examples of what might result:

1. A simple account of a street accident becomes a 'death trap claims new victim' story where the reporter had failed to link the accident with previous ones at the same spot.
2. An important job appointment and a story about bequests in a will are brought alive with new angles through the eagle eye of the night editor following up obscure details that seemed to call for explanation.
3. A planning decision of the local council is found to have vital implications for the community that were not immediately apparent in the report of a council meeting.
4. A story about an old lady found dead in a flat in which she lived alone is built up and rewritten to exemplify a campaign the newspaper is running about the effects of urban loneliness.

Angling as demonstrated in these examples shows how the creative assessment of a story can result in a recasting of the material by the subeditor and the working in of new copy.

There is also the more general angling of stories to suit the readership. Political stories, as we have seen, are sometimes presented in regional and town papers in terms of the local connections of the politician or the likely local effects of proposed legislation. A paper's political writers are aware of this and it does not often fall to the subeditor to re-nose their copy.

Other jobs can entail a considerable amount of work to bring out the local angle. The casualty figures in a crash or disaster story might include local people and the story will be edited from that viewpoint. Trade and export stories might be angled to bring out their connection with local industry and jobs.

Multiple copy sources

Sometimes a subeditor has to handle a story which has copy from more than one source so that the editing becomes a merging of several texts into one story. It might be a story with several ends in which a main section by a staff reporter has legwork provided by correspondents.

For instance an air crash might have local interviews, or a rescue story might have material from several areas. Sometimes agency copy, either from a home agency such as the Press Association or an international agency, has coverage part of which is relevant.

Some stories have foreign ends which have important links with the main story. File background has to be worked in and perhaps a financial angle from the City correspondent.

Keeping up with this amount of copy in a complicated, many-sided story requires an eye for vital detail, especially when extra copy arrives during the subbing of the story. The subeditor cannot afford to miss anything of importance. Rewriting is often the answer.

Most newspapers prefer staff reporter's copy to be used where possible, and agency copy to fill in gaps, but the agency version might provide an important element of the story. Such stories should carry a combined staff and agency credit.

Bad copy

A more obvious candidate for rewriting is the story that is badly written, either for lack of time by the reporter or through a wrongly-judged style or approach. Such copy might be sent back for the reporter to try again, but if time is short the subeditor will be instructed to 'knock it into shape'. This could include 'cleaning it up', which means taking out explicit sex, obscenities, excessive violence or things likely to offend the readers; 'sharpening it up', getting rid of dull writing, tedious phrasing and wordiness; and 'toning it down', ridding it of extravagant or emotive language, excessive use of adjectives or gush.

Some reporters who are employed because they are brilliant news gatherers are indifferent with the pen and rely upon the subeditor for the final polish. In American practice this need is recognized by the use of rewrite people separate from subeditors. In British practice it is found more useful, for page planning purposes, to keep rewriting within the subediting and production orbit, although under computerized systems more use is being made of newsroom rewrite staff. The precise boundary in rewriting between what should be done in the newsroom before submitting copy and what should be left to the subeditors is a matter for house practice. There is no doubt that basic newsroom collating of source material simplifies the subeditor's job, yet with imminent production deadlines the arrival of late or additional material is best dealt with by the subeditor who is closer to the space requirements, and is already well into the editing of the story. Daily newspapers, especially town evenings, are much more subeditors' papers than they are writers'.

Electronic aids

Electronic editing, if used properly, should offer no insoluble handicap to editing jobs that involve a degree of collating and rewriting. The computer has a prodigious capacity for the storage and retrieval of material, and though split screen allows only two stories to be displayed at a time, the main edited text can be displayed on one and material from successive other sources worked in from the other. Equally, rewriting can be done on one screen for insertion in the adjoining edited text.

In the case of the more complex stories the copy-taster can help by filtering through to the subeditor only essential material so that the subeditor does not have to do a secondary tasting job on an increasing flow of copy. If there is any need to go back to original copy later it can be stored intact in the computer, or left in 'note' form in the edited version of the story so that it is there to be seen when called up on the screen, even after typesetting.

Rewrite subbing from complex sources is a job for an experienced subeditor. Such a person can sift through a varied input of copy, do the necessary fact checking by reference book or cuttings, and within minutes begin writing the story from intro onwards – or with the intro to follow – while the copy is being typeset a section at a time. On an evening paper, especially in the subbing of the splash or late stories on page one and page two, this is a valuable facility.

The following points should be watched with rewrites, particularly by subs new to the table:

1 Is the rewrite necessary? Is using part of the reporter's copy, with subbing adjustments, just as quick or quicker?
2 Have quotes, names, ages, addresses, etc. been checked against the original copy?
3 Have quotes been transcribed accurately or paraphrased fairly?
4 Has the style of the paper been followed in spellings, abbreviations, use of colloquialisms, etc.?
5 Where subbing is on hard copy, have hand-written names been capitalized and figures written clearly so that the keyboard operator will understand them, and has a note been kept of folio numbers, typesetting and measure?

REVISING AND EDITIONIZING

National dailies in Britain publish usually in three or four editions during a production cycle lasting from about midday until 2.30 a.m. The first edition, for the circulation area furthest away, has a press time of about 6 p.m. or 7 p.m., with later editions printing at two or three-hour intervals up to about 12.30 a.m. to occupy a printing run finishing about 3 a.m. In dual centre production (say, London and Manchester), edition times are later since distances and travelling time for edition areas are shorter. This means that the first edition can contain later news, which is especially advantageous with regional sports fixtures.

Area changes in content in national paper editions are mainly concerned with sport, with full-length match reports in football and cricket being substituted on the sports pages for each readership area. One or two pages might be kept open for regional news, but the bulk of the edition news changes concern the updating of national stories in the light of later information or the substituting of later news and pictures. Stories that are likely to need updating are usually carried on Page One, the back page and on certain 'edition' inside pages, with a number of pages (other than regional 'change' pages) remaining the same through all editions.

In the case of town evening papers and local weeklies the editions are arranged to give a much greater stress to area and community news, sometimes to the extent of varying the title of the paper to emphasize its local identity. Thus the *Hoylake News and Advertiser* becomes the *Heswall and Neston News and Advertiser*, and the *Malton Gazette and Herald* becomes the *Pickering Gazette and Herald*.

Some big town evening papers run as many as six or seven editions in a production cycle lasting from about 11.30 a.m. to 5.30 p.m. In both town evenings and weeklies the pagination is arranged (depending on whether the paper is tabloid or broadsheet) so that a number of pages of area news can be replaced at each edition change. General news is updated or changed as the day goes on to fit in with the various edition times. The front and back pages change with every edition and contain the stories most likely to need updating. With tabloid-sized pages, the subeditor would need to remember that pages are made up in pairs.

In cases of great urgency it is sometimes possible to 'slip' one page (or pair of pages) on its own between editions in order to change or update a story, thus creating a special or 'slip' edition. This is done if new facts or news have to be carried and the editor does not wish to wait until the next edition to get at the page.

In the case of edition area coverage the stories are marked with their edition when subbed, and are kept ready in type to be used as the edition changes become due. Most changes on area pages involve stories destined for one edition only, but some stories are edited in different versions to different deadlines to bring out local angles for the different editions.

A story in one edition about a robbery in Bradford might be angled in another as 'Huddersfield man accused'. A story about a Bristol bride can turn up in another edition as a story about a Swindon bridegroom.

It is usual on a big evening paper to list the editions and their individual page press times on an information sheet so that subeditors, especially new ones, can refer quickly to the deadline for whatever story or edition they are working on.

Rejigs

The revised form of a story is called a *rejig*, or a *redress*. The subeditor, either on proof or screen, works in later material to update the story for the next edition. *Add* matter will go conveniently at the end, and can be subbed and typeset as a separate item to be picked up when the page is next made up. If the length allowed to the story is the same, the subeditor would have to offer a 'cut' in the original story to allow for the new material. This is the simplest form of revision, with the headline and main body of the story remaining the same.

Where the later material is more important it can involve re-nosing and re-headlining the story for the next edition. This might involve reworking the story, beginning with the latest material and working in parts of the original story that are needed, depending on space and relevance. The copy is thus treated as a new story and given a new catch-line, and is re-entered into the system as new, the original story being erased. Unless more space has been allowed for it – maybe a page lead instead of a single-column top – the subeditor must be careful to cast it off to the same length as the version it replaces. Late rejigs that turn up too long on edition times can cause serious delays by jamming up the typesetter, unless the cuts are straightforward.

Late material can also be sent out for setting as *inserts* and marked into the story as 'insert A' or 'insert B', with cuts indicated so that the same length is preserved.

For the more complex problems of revision and edition changes, see Chapter 11.

CAPTIONS

Captions, the writing of which is part of the subediting function, are a world on their own. While a picture cannot stand without them, they should not stress the obvious but should try to extend what the picture they cover is saying by offering explanation and context. For instance, a caption to a picture of a man playing a golf shot would be banal if it said:

Trevino playing a long shot

It would be better if it told the reader:

> Trevino: a record round
> or
> The Trevino style – today's picture

which at least makes the point that the picture is hot from the camera. Likewise a rail crash picture would have to say something more creative than:

> The crash scene at Extown Junction

For instance it could inform the reader:

> The upended coach in which ten survived.

Even stock pictures in TV programmes can be brought to life and not give just names. For example,

> BRUCE FORSYTH: a comedy comeback on BBC 1 (7 p.m.)

at least gives readers a vital piece of information which sends them looking into the programmes. And note the use of the name in capitals. This is a good ploy with all captions.

All these examples extend readers' awareness of the accompanying text as well as explaining the picture and, by giving a titbit of information, persuade them to read on.

While identification is paramount in a caption, time and location are not always necessary. There are also a few do's and don'ts about caption writing that can be rumbled by examining newspaper pages. For instance:

- The convention of writing a caption in the present tense is well founded and worth sticking with. It can give even a stock head shot immediacy.
- They should be adjacent to the picture and preferably under it. A caption that has to be looked for has failed.
- Captions that cover a number of pictures, containing words such as below left, above, far right, etc., can exasperate the reader by driving him or her to search around to find which is which – especially if one of the references proves to be wrong.

Most captions appear alongside the story which the picture illustrates and therefore do not have to say very much other than to justify the picture, but some display pictures do not tie in with stories and depend on their own self-contained caption. The material usually still has a news point, even if it is just to announce that this is the newly issued official portrait of one of the Royal Family, or that this is the mayoress modelling her new chain of office.

A difficulty can arise in the popular tabloids, however, where a girlie picture appears on Page Three because this is what the readers expect to find there, rather than because there is any other justification. It is here, in

184   *Modern Newspaper Editing and Production*

providing self-contained captions to primarily display pictures that the special art of the caption writer reaches its highest fulfilment; when, absolved almost from dealing in facts, the writer can use comment, whimsy and clever writing to bring the picture alive. Even so, the conventions of identification, justification and the use of the present tense should be applied. The reader has to be guided.

Caption typography

This is best kept simple. It should be different from the adjacent reading text, usually in bolder type and a size bigger. In captions of more than one line the writer should try to aim at even lines, giving instructions to square or equalize the lines. It is usually sufficient to contain the lines in a ruled 'bucket' if emphasis is needed, and the use of stars, blobs, squares and other graphic decorations to draw attention to captions should not be overdone. There are cases in which captions at the side of a picture can be set ranged against the picture and ragged on the outside.

Some captions, where the material is self-contained, and even those used with a story where the picture content is exceptional, can be usefully given a headline. It is here that a label headline, usually banished from the news stories, can be effectively used to draw the readers' attention to the message in the picture, especially if there is already a main headline on the story.

For example, a sports caption headline might say:

HOWZAT!

or

YORKED!

or

GLORY NIGHT

– the latter in the case of a Cup Final celebration.

On human interest pictures of strong content a headline might say:

THE GIRL HE
COULDN'T FORGET

or on a rescue picture:

THE MOMENT
OF FREEDOM

Picture headlines, which sometimes consist of clusters of words perhaps in the form of a quotation, are not trying to tell you the essentials of the story. These are already given in the news headline close by. They are trying to connect you with the fleeting moment in time caught by the picture. They are a rare example, as captions themselves sometimes are, of where comment is allowed to intrude into a news page, though intrude is not the right word. What the comment really does is to orchestrate and heighten what the picture is saying so that you are drawn to read the story.

CONTENTS BILLS

One of the lesser-known jobs associated with subediting is the writing of news bills, usually called *contents bills* (*posters* in the US). Part of a newspaper's self-publicity is to display these bills outside news-stands and at strategic points in the circulation area to advertise its contents.

There are various types. *Stock bills* are those that remind readers that racing form cards or television programmes or 'latest sports news' can be found in the paper that night or that morning. These are usually pre-printed. *General bills* are those that advertise an important story, and are placed at all display points in the circulation area. Thirdly, there are *local bills*, which emphasize the local connection of a story and which are placed in their own area.

On some papers a senior subeditor writes all the bills, taking the material from page proofs as each page is made ready. In others, subeditors are asked to write off bills for stories that are considered 'billable' while they subedit them.

The technique of writing a contents bill is to disclose enough information to arouse the reader's interest without saying so much that the reader has no need to buy the paper – a legitimate advertising and publicity ruse. In this sense they are not explicit as headlines are, which are intended for the reader who has already bought the paper. For instance

> FLEET IN
> STAND-BY
> SENSATION

has urgency yet suggests an intriguing range of possibilities, and is typical wording for a general bill.

> PRINCE IN
> MARRIAGE
> RIDDLE

likewise gives in general terms a story of wide-ranging possibilities, as does

> MINERS'
> STRIKE
> DECISION

It would be self-defeating (and bad practice), however, if the wording deliberately misled the reader – for example, if the fleet turned out to be the Peruvian fleet or the prince a member of an obscure royal family. A good bill has to honestly justify the 'bait' it places before the passing reader.

With local bills the place name is all-important. A national paper which avoided the town's name in a headline would bill a story to its area as HALIFAX WIFE IN DEATH RIDDLE if it felt the story were good enough to garner a few extra local sales. A town evening paper could use surburban references. The *Newcastle Evening Chronicle*, for example, might bill the same story as BLAYDON WIFE IN COURT RUMPUS and WIFE IN NEWCASTLE COURT RUMPUS. In cases of casualty figures or local election results, one story might yield a handful of local bills, each containing a different area name.

Bills can be 'local' in a different sense than in place names. A story about universities might produce bills for all university towns, or a council decision about housing be billed to all housing estates in an area. A story about theatre in general could be billed outside all London West End theatres.

Bills are unashamedly labels. While the active verb is not excluded, it is not vital. The wording thrives on evocative but very general terms like *rumpus*, *riddle*, *sensation*, *row* and *drama*, which would be frowned upon in headlines by some chief subeditors, but whose lack of precision is a virtue for the purpose of a bill. Long words should be avoided so that the words can be given maximum size for distant reading, with four or five words the maximum. The message should be simple, consisting of one thought. Bills that try to say too much fail in their purpose.

The usual system is that bills are sent, handwritten, to the circulation department who produce them by stencil or heavy hand lettering on standard headed sheets for distribution to their sites.

11 THE RUNNING STORY

A running story is one that continues to yield copy, edition by edition, throughout the day and sometimes through several days. It involves a continuous process of editing, revising and re-headlining which can occupy the subeditor up to the last press time of the last edition. It can frequently result in follow-up stories the next day with further revising through editions.

The handling of running stories is usually a senior subeditor's job because of its complexity, although a modest three- or four-paragraph story given to the newest subeditor on the table can turn into one unexpectedly and provide a career baptism of fire.

A wide variety of news events can become running stories. The more obvious ones are disasters, rescue stories with the long-running drama of the hunt for survivors; elections, with the cliffhanger of the shifting balance of parties as results are announced, and important trials taking several days in which evidence is covered by relays of reporters. Public inquiries, tennis tournaments, cricket Test matches and political crises are other stories of this sort.

Editing a running story, sometimes in the intervals between handling other stories, needs a subeditor with a cool head, an acute news sense and speed. The story has to be given shape for the first edition using the copy available, with the intro and leading paragraphs being typeset last so as to lead with the most up-to-date situation. Yet no sooner is the story complete than it is being re-shaped with later material for the next edition, with again the intro being the last part to be written and typeset.

Direct input systems, with the easy revision made possible in screen editing, can make for quick changes in all but the most complex of running stories. Here, newsroom-collated new leads can be a help (if there is time) where there is a rapidly changing situation or a variety of copy sources. As with rewrites in general, the responsibility for handling running stories falls mainly on the subeditor, with the newsroom often supplying back-up to agency inputs.

A story can be rejigged and updated as many times as there are editions if the input of copy warrants it. In the text, figures and amounts have to be watched so that the middle of the story and the headline are updated along with the intro and there is no clash or discrepancy. Later information in a running story can change the picture entirely and the whole sequence of paragraphs has to be kept under review.

For instance, in a train crash story the reader should not be told in paragraph ten that rescuers are searching for someone who in paragraph two, has been found. More precise information later can alter impressions drawn from first eye-witness accounts. Aspects of the earlier account are retained only if they are still relevant to the narrative and not in contradiction to the later facts as a fuller picture emerges. Sometimes the story as finally revised contains almost nothing of the first version.

It is important, during these revisions, that any earlier material re-incorporated at the last minute to fill out a revised story to fit the space allocated should be checked in case it is wrong or contradictory.

HANDLING THE TEXT

Here is how a typical broadsheet provincial evening paper using hard copy subbing (for ease of explanation), photo-typesetting and paste-up page make-up would handle a running story. It is based on a number of real-life actualities of the sort of problems involved:

10.00 a.m. A Reuter flash reports: SCHEDULED FLIGHT FROM LONDON AIRPORT TO CHILE TWO HOURS OVERDUE AT SANTIAGO. The newsroom checks out flights to Santiago and identifies the aircraft as almost certainly British Airways Flight Number 004, but can get no comment from the London airport authorities.

11.00 a.m. The editor at his main production conference takes a chance and decides to get a story into Page One of the first edition, due to go to press at 12.30 hours, and gives it provisionally the right-hand end column on the page, the usual late news spot, on the assumption that more copy will arrive.

11.15 a.m. A second Reuter flash reports: WRECKAGE OF AIRLINER SIGHTED BY MILITARY AIRCRAFT ON MOUNTAIN TOP NEAR CUZCO, PERU. It is 6 o'clock in the morning in Peru.

11.30 a.m. The two Reuter flashes and a background piece from the newsroom who have identified Flight 004 as a Boeing 747 are

	given to a subeditor to prepare a quick Page One story with the copy-to-printer deadline of 11.45 hours.
11.35 a.m.	The story is switched by the editor from a single-column to a double-column shape alongside an 'own correspondent' political splash and is given a headline in three lines of 60 pt caps saying:

> BRITONS FEARED
> LOST IN ANDES
> AIR DISASTER

	There is some discussion over the words 'Britons' and 'disaster' but the editor takes the view that the flight, since it is a British Airways flight from London, must be carrying some Britons and that an airliner coming down on a mountain top must come in the disaster category. He is still taking a chance, however, since it is not known for certain that there are Britons on board, nor has the airline confirmed that it is the Flight 004 Boeing 747, though they have not denied that it is (Figure 29).
11.45 a.m.	The subeditor has managed to put together seven paragraphs based on the Reuter flashes, the newsroom inquiries, information on the aircraft's known stops and help from an atlas, and the story is keyboarded in for typesetting a folio at a time as soon as each is written. A third Reuter flash, HELICOPTER REPORTS NO SIGN OF LIFE IN AIRLINER WRECKAGE NEAR CUZCO, arrives just in time for the intro paragraph, which the subeditor has kept till last, and he is able to re-word it:

> A number of Britons were believed to be on board a British Airways Boeing 747 jet which crashed on an Andes mountain top in Peru early today with a feared high loss of life.

11.49 a.m.	The re-worded intro paragraph is keyboarded in for typesetting a tolerable four minutes late. The earlier pages have gone well and the editor hopes still to have the edition out on time.
12.15 p.m.	A 'lead' story now comes from the agency, building up the earlier flashes and 'rush' message about the flight being off course, together with the agency's own background sources, into a twelve-paragraph story, but it goes no further than the subeditor's own effort. There have been no further rushes and the editor is pleased there is a useful story in the first edition.
12.30 p.m.	A news agency rush, followed by snaps giving details, announces that twenty-seven Britons joined Flight 004 at

	Heathrow. British Airways say that first indications suggest there is a heavy loss of life, and that Peruvian Army helicopter units are attempting to reach the wreckage, which is on the edge of a small plateau. The picture library offers recent stock pictures of British Airways Boeing 747s.
12.40 p.m.	The editor consults with the managing editor and the chief subeditor and decides to make the story the Page One splash now that twenty-seven Britons are known to be involved. The managing editor orders a location map from the retouching artist, who doubles as graphics artist, and gives him the atlas reference and the geographical details that are known. The chief subeditor is asked to reduce the old political splash to a double column in the last two columns.
12.45 p.m.	Further agency snaps arrive. The Peruvian Army is being helpful. It is airlifting a rescue team with radio by helicopter to the crash scene. The team is in contact with its airbase near Cuzco.
12.50 p.m.	The Press Association (national) agency puts out an ADD ANDES AIR CRASH STORY saying that the twenty-seven Britons on board are believed to include a national team of twelve women gymnasts bound for a pre-Olympic work-out at Valparaiso. The story gives no names. With this information the news desk checks out the likely members of the team who would have taken Flight 004. As is usual, British Airways do not name crash victims until the next-of-kin have been informed.
12.55 p.m.	The subeditor, armed with a proof of the first edition story, the Reuter lead, the PA 'add' story, the latest snaps and a piece from the newsroom on the women gymnasts, starts updating the story as the second edition splash with a 1.15 copy deadline.

The editor soothes the political correspondent for downgrading his important story, which is to do with the effect of new housing legislation on the town, and suggests a follow-up feature on it for the next day's paper, in addition to the half lead still on Page One.

There are 20 minutes in which to prepare the splash. Twelve column inches (300 mm) plus an intro across three columns are being kept for it. The centre and top of the page are redrawn to allow for the display the editor wants, and the column one story about a local councillor finishes down column six. The political lead and the tie-in material above it are subbed down to fit the space in columns seven and eight.

The running story 191

Figure 29 *First inkling of a disaster story – layout 1*

Figure 30 *The story develops – layout 2*

Pictures are now the real problem, the stock jumbo jet picture being a bit static (Figure 30).

The first edition air crash story, which was set in 10 pt across two columns, is thrown away since the new story is mostly in single column and is almost a complete rewrite to take in the new facts. Three quarters of the splash leaves the subeditor as handwritten copy to go for typesetting.

The chief subeditor writes the new headline which says, in 120 pt 'disaster' caps:

> 27 BRITONS
> FEARED DEAD IN
> JET DISASTER

It carries a strap line:

> Big toll as London airliner hits mountain top

The story is wrapped round a three-column picture of the Boeing 747 from stock, carefully captioned. To the left of the intro is the map giving the approximate crash position and a dotted line showing the estimated flight path, and also Cuzco, the nearest city. It is the best that can be done in illustration since the difficulties of the site, despite the extra five hours available in western South America, will make crash pictures for any of the afternoon's editions unlikely.

1.05 p.m. Press Association puts out a statement from British Airways saying that the aircraft *is* Flight 004 and that the jumbo jet has 220 people on board, well under its capacity load, including a crew of ten, among them sixty-seven Britons - not twenty-seven – and that there is no information about possible survivors. It is thought that further Britons might have boarded the aircraft at the stop at New York. The statement gives the names of the captain and first officer but no other names.

The new figures of sixty-seven Britons is substituted in the headline and intro and the new information worked into the story, partly as the intro and partly as an insert to bring it well up in the story, which is now half set. The subeditor is still handicapped by the lack of precise information about what has happened.

1.15 p.m. The intro paragraph is finally keyboarded in. It takes all the information from agencies and newsroom, together with an end piece about previous jumbo jet crashes, to fill the allocated 12 inches. The chief subeditor offers a brief compliment to the sub who has tied it all together. It is all in a day's work and praise is not easily come by on subs' tables.

1.30 p.m. Further snaps confirm that the women gymnasts' team is on board, and say that no word had been received from the pilot by the control tower at Santiago saying that anything was amiss, despite the fact that the aircraft appeared to have been 200 miles off course. The editor discusses the situation with the managing editor and the chief subeditor and decides to stay with the story as the splash since, despite problems, it continues to be the most important live story of the day and a good one for the evening papers.

Also, with the political story and its local connections as the half-lead, the page now has good balance, though it could badly do with a live picture. The chief subeditor suggests using an attractive picture showing a boy with a dog he had rescued from a culvert as a useful mid-page picture and story on Page One. The editor agrees that this is a good idea since it will help to give the page an air of busyness that a town broadsheet needs.

1.40 p.m. The group's aeronautical correspondent, based in London, weighs in with a comment piece about instrument failure, due to electrical disturbances in the atmosphere, being the likely cause of the aircraft being off course. He has checked the weather reports and found that, though not in a storm, the aircraft could have been affected by violent electrical disturbances which had been reported between the Pacific and this part of the Andes. The copy is passed to the subeditor with the instruction to 'bring it up' in the story.

1.43 p.m. A Reuter rush message says: PERUVIAN ARMY RESCUE TEAM IN RADIO MESSAGE TO BASE REPORT TWO PEOPLE FOUND ALIVE IN CRASHED JUMBO JET.

1.45 p.m. The managing editor claws back a further 10 inches of space on Page One for the 3 o'clock edition by moving the councillor story on to Page Two (Figure 31) which goes to press 15 minutes before Page One. The air crash story, with its stock picture, map, headline display and text is now taking up nearly half the page. The copy on previous jumbo jet crashes is hived off from the main piece and given to another subeditor to prepare with a separate headline as a tie-in piece to the splash, placed under the map in columns one and two.

1.50 p.m. Further snaps arrive from Reuter describing the wreckage, followed by a new lead re-nosing the story on the two people found alive and describing the search and rescue operation. In a note to news editors, Reuters say that their man in Santiago has now reached Cuzco and linked up with the stringer there and has found the Army authorities helpful. The Army has even agreed to airlift them both to the forward base from which the rescue is being directed.

Meanwhile the newsroom has produced a background story about the women gymnasts with a recent picture of the team, two of whom come from the circulation area, though it is not known if all of them boarded the flight.

There is some discussion about this between the editor and the news editor. The picture would be useful on Page One but with the names of the passengers still not released, it poses a delicate problem. The editor decides, on the ground of taste, not to use the picture for the present.

2.00 p.m. The editor changes his mind about the picture and decides it will be reasonable to use it provided it has a non-committal caption and goes with a separate story under its own headline. It must not be interpreted in any way as being part of a list of casualties.

The managing editor, who has more or less taken over control of the crash situation, asks the chief subeditor to find a home for the projected dog rescue story on Page Two (the usual home for stories taken off Page One), so as to clear a way for the women gymnasts. The picture is schemed across four columns below the fold, almost the same shape as the dog picture was going to be (slightly shallower, in fact) and the subeditor who had done the jumbo jet tie-in story is given the job of preparing careful copy to go with the picture. This is allocated to columns one and two in place of the jumbo tie-in. The newsroom has simply supplied indentification material, not being certain how it would be used with the main crash story, if at all.

2.05 p.m. The subeditor begins revising the splash for the 2.30 p.m. copy deadline for the 3 o'clock edition and marks the first folio of handwritten copy *intro three paras to come*, and then works in the new details, picking up the old text from a print-out, where it still applies, and restoring the background material on previous jumbo jet crashes, which is now being run on at the end of the splash. This will make a useful last-minute cut at the make-up stage if the story is over-set since it is the least important part of it.

2.22 p.m. The Reuters team who have managed to get their first message to London from the forward base, using Army facilities, send a description of the rescue scene on the plateau, where it is now 9 a.m. local time. Cloud has lifted and twelve rescuers are combing the wreckage, though hampered by the high altitude. The subeditor decides to work this material into the intro since the main story is now two-thirds into the photosetter.

2.30 p.m. A Reuters rush says: THREE CRASH SURVIVORS FOUND, ENGLISH STEWARDESS AND TWO PASSENGERS IN SERIOUS CONDITION. NO SIGN FURTHER SURVIVORS. The information is just in time to be included in the intro, which now reads:

> An English air stewardess and two passengers are thought to be the only survivors of the British Airways Boeing 747 jumbo jet from London which crashed in the high Andes in Peru early today.
>
> The stewardess, who is only slightly hurt, was being airlifted by helicopter along with two injured passengers, to a hospital at Cuzco, 200 miles from the crash scene.
>
> Peruvian Army mountain rescue troops are still

The running story 195

searching the wreckage of the airliner whose 220 passengers and crew include 67 Britons. Among them are members of a team of women gymnasts who boarded the aircraft in London. Two of the team are believed to come from Lancashire.

This means leaving out the rescue colour material for the present, but the subeditor knows it is important to nose the story on the latest information since the fact of the crash will be already known to readers through radio and television newscasts. The thrust of the story must now centre around the rescue operation, in which the news agency team has an unbeatable advantage by getting so quickly to the forward base. The one disadvantage of this is that there will be no exclusive angles available to any paper.

The final step for the 3 o'clock edition is to update the splash headlines, which are changed to read:

Air hostess and couple survive out of 220 in air disaster
THREE ALIVE IN BRITISH JUMBO JET WRECK-AGE

The splash story now spills round the headline and picture of the gymnasts, running down almost to the foot of column size (see layout sketch three).

2.50 p.m. A Press Association 'with crash' story begins running, giving the names of all the crew of the airliner, which have been released by British Airways. Three stewardesses are included, but the one that has been found alive is not identified. The PA story promises pictures to follow.

3.00 p.m. The editor holds the final Page One conference with senior colleagues before going off duty just before 4 o'clock and handing over to the managing editor. They decide to build up the local end – the two Lancashire women in the team of gymnasts – but to keep the splash in its present space. In the absence of any pictures from the site, the picture of the gymnasts, which has been provided by a reader, becomes a key factor in the Page One projection. Also, it is exclusive.

The news editor suggests sending a reporter to see the families of the two women, but the editor rules against this on the ground of taste, since it has not been confirmed that they actually caught the flight. Also, the official casualty figures have still not been released, nor are they likely to be before the last press time at 4.45 p.m. For the present, the local end would have to rest on the team picture and the background material from cuttings which the newsroom has researched.

Figure 31 *It takes more. . .layout 3* **Figure 32** *. . .and more of the page – layout 4*

	The editor suggests that the local end should be included in the strap line.
3.10 p.m.	The subeditor, meanwhile, begins revising the story for the 3.30 p.m. copy deadline for the 4 o'clock edition, lifting the reference to the two Lancashire gymnasts, on instructions, into the second paragraph. He covers this by saying that although the names of the crash victims have yet to be released, the two were members of the team of gymnasts who were known to have caught the flight in London.
3.25 p.m.	A Reuters rush gives the names of the survivors announced by the hospital in Cuzco, to which they have been taken. The stewardess is Alice Merrin, aged 26, of Hampstead, London, and the other two are a husband and wife called Williams, from Cardiff.

The subeditor quickly re-noses the splash, on which he is working, to bring in the new information, dropping down the local reference to the gymnasts to paragraph four. By this time the promised Press Association pictures of crew members have arrived and a stand-by two-column picture of Alice

Merrin is got ready for the page. The managing editor alters the lay-out to make way for it (Figure 32).

This is done rather cleverly. The pictures of the aircraft, the gymnasts team and the map are moved round, the map finishing up down columns five and six, so that most of the elements of the page, including the rejigged splash can still be used. The main splash head is broken up into four lines with the same type and wording so that the two-column pictures of the air stewardess can be placed at the top of columns five and six. The gymnasts copy is cut and reset into a short single column piece in column one, a story is lost down column three, and the rest of the material on the page is lifted as it stands.

3.37 p.m. An agency second new lead noses on the survivors' names and also gives an eyewitness account of the crash scene as seen by their correspondents from an Army helicopter. This enterprising piece of reportage comes too late for the revised story, which is past its press time for the 4 o'clock edition. The managing editor decides to get back at the text in the 4.45 'slip' page, which is available for mopping up late stories on the front page, and late racing and sports fixtures on the back. The practice here is that the page plan is not altered but that quick in-and-out text changes can be made or the odd headline changed. The copy-to-typesetter deadline for the 'slip' page is 4.20 p.m.

3.50 p.m. Reuter snaps report the air stewardess as saying:

> AIRCRAFT RAN INTO VIOLENT THUNDER, LIGHTNING BEFORE CRASH. PILOT UNABLE TO REACH AIRPORT CONTROL AT LIMA.

This first inkling of the crash cause from a survivor is worth a new intro, and also the strap line is changed to read:

Air girl Alice tells of the last minutes before disaster

The first eight paragraphs of the story are reworked with the local reference dropping down further to paragraph six to allow in the later news. Some of the rescue description has to be left out so that the reworked paragraphs fit the space on the page. The local end is still covered in the separate background piece now above the picture of the gymnasts, but there have been no further developments to this part of the story and the decision to nose the whole splash on the survivors is inevitable. Also, the stewardess's story bears out the perceptive item of the group's aeronautical correspondent about local weather conditions.

4.15 p.m. An agency flash says: PILOT'S BLACK BOX RECORDER RECOVERED BY HELICOPTER TEAM. Since this piece of equipment will hold the secrets of the aircraft's last minutes, the chief subeditor prepares it, despite its lateness, as a separate bold cut-off to drop into the splash story under a simple headline of late news flash. The managing editor agrees that this is the best way to handle it, and a cut is made in the text to accommodate it. It will stand out in the page to achieve its purpose.

This last dramatic intervention in the story marks the end of the evening paper's coverage of the air crash. With local time five hours behind Britain, there should be plenty of material for a morning paper follow-up, with a detailed interview with the rescued stewardess being the prize to aim for, and probably crash pictures and more technical data about the causes. By the end of the morning papers' edition run there is every likelihood, too, that an official casualty list will have been issued. It could be as busy a night on the morning paper as it has been a busy day on the evening paper.

- While the model chosen in this chapter is of a hard copy subbing operation feeding into printer-operated photosetters, the routine would be the same with direct input of copy and screen editing. If anything, there would be a gain in speed in taking in late amendment of copy on screen, either before or during photosetting.

12 PUTTING IT ALL TOGETHER

The final stage of editorial production is the putting together of the stories and pictures that have been chosen and edited so that they form a newspaper – in other words, the making up of the pages.

The pages still have to be turned into printing plates, nowadays nearly all of polymer, to be attached to the presses which print the newspaper, but it is with page make-up, which takes place in the composing room, that editorial involvement in the production process reaches its peak and comes to an end.

Nowhere is the cold type revolution more apparent than in the modern composing room. Instead of the long lines of heavy metal-topped 'stones' on which the type was assembled in chases from its galleys, there are banks of easels or frames at which the page compositor stands to make up the pages. The 'chases' are graphed cards marked vertically in column widths and horizontally in centimetres and millimetres, now universally used to define picture and story depths in page layouts. The type consists of sheets of photographic bromide paper being delivered from the photosetters into baskets, and bearing text and headlines exactly as they will look in the newspaper. These the compositor cuts up and pastes on to the page.

Yet the care with which the page ingredients are put together, and the editorial control over content and spacing, remain the same. The pages have to be got ready to match the layouts, and checked and passed by the lawyer, or legal reader, the back bench executive involved, and by the editor, before being photographed to produce the film transparency from which the printing plate will be made.

COMPOSING AREA

The switch to cold type makes possible a more integrated floor layout which brings the editorial and make-up functions closer together physically, and requires only a fraction of the space taken previously. The banks of Linotype machines, cases of headline type and the proof-pulling machines are replaced by photosetters which are about twice the size of a large domestic freezer, and of which there are usually two or three on a small or medium-sized paper and four or five on a big national paper.

Copy that has been keyboarded and processed through the computer can be delivered by a photosetter in the type and measure required at speeds of more than a thousand lines a minute, compared to five or six lines a minute on a manually operated line-caster. If the story has to be rejigged by the subeditor it can be called up from the computer, updated and altered, and be re-run through the photosetter in a matter of minutes.

A one-floor layout in which the editorial and composing area are separated by the computer and the photosetters is ideal. The composing area should nevertheless be designed with care to achieve a controllable work-flow, particularly if more than one newspaper is being set in type on the same equipment (Figure 33).

Type, once setting commences, drops with great speed and should be identified from its header instructions, logged and carried to the pages with a fail-safe system that minimizes the possibility of losing things. The system requires a well organized 'random', as with hot metal type, which is controlled by a random supervisor who routes type to the pages where it is awaited, as it used to be when carried in galleys. Despite the switch to photographically generated type, the sequence of typesetting and make-up remains the same.

The random requires baskets with type 'flags' and a type log book to keep account of what has been received and what is still awaited. It is essential that at the keying stage stories are properly identified with catch line, page and edition number and source by subeditors so that queries can be quickly settled and re-keying minimized.

There are good opportunities for systems people to plan a composing area in which typesetting, random and page frames are placed in logical relationship to each other so that work movements are minimized and time properly ordered. In addition to coping with typesetting, it needs to be alongside the cameras and darkroom producing bromides of pictures and advertisements needed for the pages. These facilities, in their turn, should be in easy reach of the art desk or editorial and advertising production desks where picture and artwork instructions originate. If keyboarding of copy is done by separate operators, and not by the subeditors, then the space for their VDU work-stations has to be allowed for as well.

At the other end of the work chain should be the page copiers that produce 'proofs' of the pages for final checking, and the camera room that makes the page transparency from which the printing plate will be derived.

It is essential that the composing area is effectively managed so that staff

Putting it all together 201

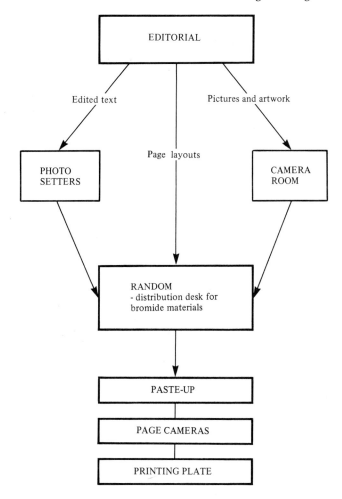

Figure 33 *An example of work-flow in a modern computerized composing room*

rostering and meal breaks are in line with page requirements, the sequence and priority of page formes is watched and 'off stone' times adhered to. Technical advances can be frustrated by organizational inefficiency.

The main problems of control arise from the rapid input of bromide material into the composing area from the photosetters delivering text and headlines (Figure 34), and the camera outputting page-ready bromides of advertisements, pictures and other artwork. Pictures and advertisements, as with type, need fetching, logging and delivering to the pages. The danger of

Figure 34 *A photosetter delivering a type bromide ready for the page*

serious muddle looms quite high, and the random system needs to allow, above all, for a clear separation of editorial and advertising material. There should be separate overseers for each who are familiar with the editorial dummy and the advertising dummy, and the various editions, and pages, and can check and chase material when necessary.

One advantage of advertisement bromides is that if they arrive over-width or over-depth, due to variations in the format and size of the papers for which they have been designed, they can be reduced on camera quickly to the size and space allocated to them on the page. Likewise they can be enlarged.

MAKE-UP

It would be logical, once text and headlines have been entered into the computer, to retain them there and make them up into a page on the screen. *Full page composition* has always been the dream of system makers and the graphics facilities now available have solved most of the problems posed by advertisement and picture material. Taking the ingredients out of the computer and piecing them together on a board in order to produce a page negative seems a waste if it could all be done inside the computer. Then the photosetter could print the entire page in one instead of the individual items.

Screen make-up, in fact, is now totally feasible – yet paste-up is the option generally chosen, even in American papers where newspaper layout is simpler and more formatted. The reason for this is its flexibility and the facility it allows for display effects unhampered by the modular patterns imposed by the computer, and also by the continued lack of flexibility in the use of computer graphics.

While the capability of printing systems continues to expand to meet problems that once raised doubts, the dream of totally flexible full page composition still eludes the designers, despite the great advances in graphics generation. It is for this reason that in the minority of offices where it is in use, full page composition invariably involves the making up of pages on screen with blanks for the prints of half-tones, and sometimes of advertisements. These are pasted on to the page before the transparency is made from which the printing plate will be derived.

Even in the US, according to William D. Rinehart, Technical Vice President of the American Newspaper Publishers' Association, in a letter to the author, more than half of the country's newspapers are not expected to have moved on to full page composition before the 1990s.

Paste-up also has advantages that suit the more complex design practices of British newspapers. Given the materials, the eye and hand can work at a rate that was never possible with hot-metal make-up, meeting at the same time the most exacting demands of layout, and achieving visual effects with type and half-tone that were not previously possible. The method provides the ultimate in flexibility for the elements comprising the page.

The speed with which pictures and artwork material can be 'shot' to produce the required bromide, and the accuracy with which stories can be cast off to length on screen by the subeditors, means that paste-up can proceed without the hold-ups associated with the hot-metal method. Once the material has begun to flow from the typesetters it can be quickly cut up and attached to the page 'forme' by a deft paste-up compositor with a scalpel (Figure 35). The forme (which consists of two pages in the case of tabloid papers) is taped to the working frames usually on its own card so that it can be kept rigid when taken to the camera after completion.

The method of pasting, using hot wax, is quick and mess-free and the text and headlines can be docked conveniently on the board alongside the page while they are being cut up to fit the pattern of the layout. The pieces of

Figure 35 *A paste-up compositor at work on a page. The type bromides are being cut up and pasted on to a page card to the layout plan on the left in the picture. At the top of the frame can be seen the various stick-on tape rules that are used*

bromide can be lifted, and re-attached to achieve final position and easily moved about to make way for inserts and rejigged material where necessary, or to place crossheads. It is useful, however, for the paste-up operation if headline, text and related by-lines, crossheads and captions are all entered into the photosetter together so that they arrive on the same piece of bromide. The position of the relative elements can then be achieved on the card in one operation.

While dividing rules to separate columns and stories can be drawn on the page by a good paste-up comp, a finer finish is achieved by using adhesive 'electronic' taped rules, which are available in 1 pt, 2 pt, 4 pt, 6 pt and 10 pt widths and in decorative finishes. Panels can likewise be quickly made up on the page with taped rules, although most modern computerized systems can deliver formatted panels which can be programmed into the photosetter. These are set with the text and headline already in position. This formatted method also makes panels with rounded corners possible.

Pictures

The bromide paper on which pictures for the page are printed after they have been cropped, scaled to size and 'shot', is the same weight as that used for headlines and text. Once carried to the page pictures it can be trimmed with a scalpel by the paste-up comp in the same way as the text so that proper spacing between items can be allowed. It is common, at the final space check, when all the elements have been married together, for pictures to be given a discreet trim where this helps the visual balance.

With the paste-up method, cut-outs and montages of linked pictures can be put together or adjusted at make-up stage with the scalpel to as good an effect as complete camera-ready artwork would have achieved, while being at the same time more flexible.

Headlines and text

Photosetters do what they are programmed to do. In a well-planned operation this means that the appropriate type masters have been fitted at the outset, and sizes of type, setting measures and line spacing formatted so that copy can be set with the minimum of commands. A photosetter can be fitted to reproduce the type style and sizes of whatever paper is using it. If it fails to do this – giving wrong line spacing and mis-matching types – then it has not been properly fitted out and formatted at the start. Re-formatting should be carried out.

Much of the body type used in hot metal operations was old and, on change-over to photosetting, the replacement of some battered old Excelsior or Ionic by a clearer more modern face resulted in an improvement in many a paper's type format, and a gain in readability, which is helped by the method of printing plate-making used in conjunction with photosetting.

The idea with photosetting is to rationalize setting requirements on to programmable keys, thus cutting down the key-stroking and reducing the margin of error. Corrections and updates, even with formatted material such as television programmes, can still be made by recalling the material and keying in the required changes without disturbing the format. Style characteristics such as drop letters, underscored crosshead and panelled by-lines can all be formatted.

In headline writing, character counts can be given by the computer, so that a headline is assured of an accurate fit before being consigned to the typesetter. Subbed stories can be H & J'd and their setting length and word count relayed on the screen before being set, so that they can be properly cast off. There is no excuse for stories arriving on the page too short or too long.

Photosetting, if properly used, should thus greatly improve the fitting and casting off of headlines and text. Trouble-free setting in type that fits should follow from accurate key-stroking, with a saving of time and work movements at the page make-up stage.

STONE SUBBING

The person who supervises page make-up, the stone-sub, is the last editorial person to read the page before it is passed by the editor or senior production executive and goes to press, and this function remains a vital one under pasted make-up.

Under the hot metal system, the stone sub stood over the stone at which the stone hand, or make-up printer, assembled the page from the lines of metal type and the metal blocks of pictures and advertisements, and made any text cuts and adjustments to bring the page into line with the layout and the editor's requirements. For this purpose, galley proofs of the material were used.

The stone sub cut stories that were overset, called up more text for stories that fell short, and adjusted the layout, where necessary, to secure a good fit and a balanced result. Where possible, cuts were made towards the end of stories where the least important facts might be expected to be, or lines saved in paragraphs by removing words near the end, the resetting thus being confined to odd lines. At the same time care was taken not to cut out a fact that supported a headline or a crosshead.

Also on galley proofs would be marked any proof readers' corrections of typesetting mistakes, or literals, made by the line-caster operator. The prime function of the proof readers – or correctors of the press to give them their full title – was to mark literals so that the lines would be reset and dropped into the pages. For this purpose they used an elaborate system of readers' marks (listed in *Hart's Rules For Compositors and Readers, 39th Ed. (Oxford University Press 1983)*) for the various types of error – letter or word deletion or substitution, wrong fount letters, missing copy, spacing faults, transposed lines or damaged type. At the same time they checked setting for correct use of house style.

The use of editing terminals with photosetting enables all text to be checked for 'setting errors' and house style before it is set so that proof readers as such are no longer needed. Either proof reading is now done on the terminals after subediting, but before typesetting, or the function becomes part of the job of the revise sub. The general tendency is for proof reading in the old-fashioned sense to be phased out, and the responsibility for a clean, accurate text to fall on the subeditor.

The use of paste-up has not diminished, however, the importance of stone-subbing. Nor, curiously, have the terms relating to it changed in most newspaper offices. The word 'stone', already outdated under hot metal since it related to the long-disused stone bench on which pages were originally made up, has shown a remarkable capacity for survival into the computer age.

The job of the stone-subeditor is precisely what it always was – to supervise the make-up of the page to the instructions on the layout and the editor's requirements. While the use of photoset material, if properly prepared, has made for accurate casting-off of copy by subeditors where they have control of the terminals, there are still reasons why stories fail to fit at the paste-up stage. Legal alterations, if made on screen after subbing, can shorten and alter stories. Inserts and add matter in running stories can rearrange the paragraph sequence and entail cuts. Layout faults or late changes in advertisements can lead to adjustments in make-up. Pictures or stories, or parts of stories, being killed for various reasons, including taste and backbench changes of mind, can lead to the page being partly rejigged on the paste-up board. Experienced production journalists know the pitfalls that yawn just when everything seems to be fitting well and pages are getting to press on time.

Even if headlines and text are as they should be, there remains the visual side of make-up – ensuring in a hand-and-eye operation that the space achieved on the board is going to result in a well-balanced page, that stories do not turn on half lines, or jack lines, at the top of each 'leg' where they run in several legs across the page.

Paste-up cutting

The art of cutting at paste-up stage is to avoid, where possible, having to reset material at a late stage since this is the time when the setting flow is at its heaviest in the photosetters. Some use of re-setting is inevitable, however, and there is usually a terminal in the composing area where the stone-sub can recall stories to alter or cut them. Rerunning the story under its catchline is the best way where there are a number of alterations to setting and sequence to incorporate. With small amounts of setting it is easier to start a new file in the computer with a new catchline so that the reset material is delivered as an entity on its own. Such items must be marked clearly to the page so that they are properly routed. On some systems they can be put into a 'priority set' queue to avoid setting delay or the story is keyed in such a way that only the corrected parts are reset.

Either method is best avoided if late cuts can be done on the page with the scalpel. Clean line or paragraph cuts offer no difficulty here, but it is possible to achieve more complicated cuts if the paste-up comp taking instructions has a deft hand. Lines can be saved by cutting and reworking words at the end of paragraphs. Pieces can even be cut out of the middle of sections and half-lines joined up.

Attaching punctuation such as close quotes or turning commas into full stops offer no difficulty to a scalpel artist. Likewise a badly spaced line can be cut up and re-spaced to instructions, or tight headlines cut and spaced out to read better. Accurately done, cut and re-spaced lines are a fast way of making things fit and look no different from straight setting to the naked eye on final print.

There must be close attention to detail, however, to avoid parts of the story disappearing under the scalpel. It is a useful safeguard to make use of line print-out proofs, or page copies, on which the stone-sub's cutting instructions can be marked for the paste-up comp. These can be preserved in case of queries. Modern line print-out facilities which deliver a proof of the story in exact replica before the bromide version leaves the photosetter, are a great improvement on earlier print-out facilities.

Spacing

The manoeuvrability of material with pasted make-up comes as a pleasant novelty to production journalists accustomed to the rigid methods of hot metal make-up. The freedom, however, can result in the spacing on the page suffering from too much variability unless care is taken. Visual balance depends as much on organized white space as it does on type.

The graphed column markings (in non-reproducing blue lines) on the paste-up cards help in this but the eye ultimately is the guide, especially where bastard measure crosses the column lines, or where panelled text is being used. The stone-sub dealing with paste-up pages will find that a lot more time is taken in judging the space between type and rules and around headlines, panels and pictures.

The printing tradition of maintaining a 3 pt space between text and column rules is worth preserving as a basic standard. Separation between items should be even across a page except where extra indent on a measure is allowed to give greater white. Line spacing of the various measures comes formatted from the typesetters. Since the screen facility allows a precise cast-off of copy to be made by the subeditor there should be little need to increase this line spacing in order to fill a given depth, unless the layout is badly calculated. Then the paragraphs can be cut and re-spaced, but this should be done with caution. The sudden use of excessive white space between paragraphs or, worse still, between lines, can give the page a rough look. It is better to call up more text where this is possible.

A clear separation of at least 6 pt should be allowed where text adjoins heavy borders or between text and advert rules. The same space should be allowed between text and headline.

Headline line spacing should be formatted in the computer to suit the size of type. Since this has to allow for ascenders and descenders, it is sometimes necessary to adjust the space where they are absent to prevent excessive linear white. Headline spacing on the whole is a matter of style and the best guide is familiarity with the paper's appearance in cases where formatted spacing looks wrong.

Space between words in a headline should be equal to the average of one character in the given type. Cutting the words to re-space them should be done with care, since a heavily whited headline draws the space problem to the reader's attention.

The space surrounding a picture should correspond with that separating headline from text. Where picture and text occur together under a headline they should line exactly visually. Where a line caption runs under a picture it should ideally be 6 pt from the bottom of the picture and there should be 12 pt between the caption and whatever comes below it.

The ease with which pictures can be trimmed on the page is one of the blessings of paste-up, but care should be taken not to trim out essential detail. A person's head can survive losing a bit of hair but to lose some chin damages the face.

While page, advertisement and picture depths are measured in centimetres and millimetres in paste-up, the use of the printing terms points, ems and ens have survived into photosetting and it is useful to measure space in this way, visualizing it in terms of 2 pt, 3 pt and 6 pt leading (pronounced *ledding*).

A fault with some newspapers produced from photoset materials lies in erratic or non-existent spacing with type and pictures jammed against rules or against each other.

There is no reason why, with care, a page derived from paste-up should look any different from one made up on hot metal. It has every possibility of looking better.

PRINTING THE PAPER

The development of hard polymer for printing surfaces since the early 1980s has resolved the problem of durability that existed with plates derived from paste-up pages. Offset printing in the early days had been bugged by plate wear and breakage. The problem of durability became more acute when newspapers (see Chapter 1) wanted to marry the cost-saving advantages of photocomposition to the continued use of their rotary presses. Today, photocomposition offers newspapers the choice of either printing by direct impression on rotary presses, using polymer relief plates, or using smooth polymer plates on the new-style, tougher web offset presses.

Figure 36 *The page camera in use. The operator is producing a film positive of the pasted-up page from which the polymer printing plate will be derived*

Figure 38 *In contrast to the previous pictures, a page is here being made up under the traditional hot metal system. Lines, or slugs, of type and metal 'blocks' of the pictures and advertisements are being put together in a frame and 'hammered' flat with a leather mallet and block. From this a mould is taken from which the curved metal printing plate, used in the traditional letterpress process, is cast*

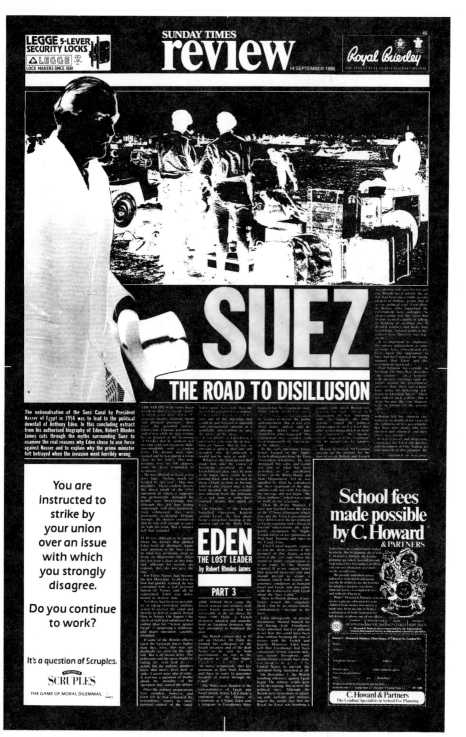

Figure 37 *A page of* The Sunday Times *at the film positive stage after being photographed. Ultra violet light is used to transfer the image in reverse on to the sensitized polymer surface of the printing plate. This will be used on the rotary presses which print* The Sunday Times

The relief plate now used for printing most of Britain's national newspapers and some of the bigger provincial ones is, in fact, a thin metal plate coated with thick hard polymer. Figure 36 shows a camera operator preparing to make a film transparency of a pasted up page – an example is *The Sunday Times* page in Figure 37 – for the purpose of using it to produce a relief printing plate. The image is 'burned' into the polymer by exposing the transparency to it under ultra-violet light. The waste polymer that has received the light is rinsed away and the rest fixed, leaving the page image in relief, and in reverse, ready for use on the rotary press.

By using this method, plate-making from photocomposed materials no longer entails delay through having to convert to metal. It is even possible, in the case of a late discovered editorial error, to 'doctor' the transparency as effectively as it was to 'bash' a word on a traditional metal stereo plate.

Figure 38 shows how a page is made up ready for moulding under the old hot metal system. Here the mould, or flong, was taken from the page under high pressure in a moulding press and was used to cast the printing plate.

Most new developments in plate-making are based on the expectation of wider use of the new web offset presses as rotary presses come up for replacement. The need to have a colour facility is a factor. Run-of-press colour printing (as opposed to pre-printed colour) requires the web offset process for the colour method by which four separate plates print in register to produce the correct colour tones. The method of making web offset plates by photo-exposure on to a smooth sensitized surface also interfaces easily with facsimile transmitted pages, in which the page signal can be uncoded and fed straight into a plate-maker without the need to produce a bromide, thus saving time. Pages made up by full screen composition can likewise be fed into a plate-maker. A variation of this is filmless plate-making, used mainly in Germany, in which the pasted-up page, on its card, is placed in a sealed frame and imaged directly on to a sensitized polymer plate for offset use.

Thus developments in plate-making are making for faster printing and easier page changes.

13 VOICES AND AUDIENCES

In examining subediting and headline writing techniques in the previous chapters we had in mind the broad area of news content. There are, of course, many sorts of newspapers with many sorts of readership, and therefore differences of content. While local and regional papers mostly cover all sections of their community, national papers tend to be read by distinctive social and educational groups. Some even have predominant age and sex readership pattern or a consciously political readership. There are also newspapers for more specific interests such as, in Britain, *The Financial Times* (business community), the *Morning Advertiser* (liquor trade), the various church and political party newspapers and those serving industrial, civic and leisure groups.

Though the news content and readership of newspapers thus varies widely, the techniques and technology used in producing them have common ground. Journalists are able to – and do – move from paper to paper across a wide spectrum of readership during their professional lives without too much trouble in adapting. Yet there are differences that a subeditor will encounter which have not been mentioned so far, though they are not marked enough to invalidate anything that has been said. Take the question of vocabulary.

SPECIAL VOCABULARIES

Anyone reading about a news event in, say, the *Western Mail*, published in Cardiff, and in *The Times*, published in London, will be struck by the similarity in structure and in the handling of the material more than by the difference. Take this political story in *The Times* with its characteristic blendings of fact and assessment as filtered through the pen of political reporter Philip Webster:

The Government expects a narrow majority in the Commons tonight for its shops Bill, which removes restrictions on Sunday trading.

As the churches and trade unions united yesterday in a final appeal to MPs to oppose the second reading, it seemed that the Government's tactic of allowing a free vote on later stages of the Bill had bought off just enough rebels to ensure its passage.

Up to 40 Tory MPs are still expected to oppose the Bill in the vote at midnight. Over the weekend whips telephoned potential rebels, commending the concessions by the Government, and explaining that they would have opportunities later to vote for compromise solutions.

Mr Ivor Stanbrook, Conservative MP for Orpington and a leading critic, said yesterday that of the 80 or so MPs who had been minded not to support the Government, about a dozen would have been 'seduced' into backing it by the ploy of free votes later.

'It is a bluff. Once the Government has got its vote, on a three-line whip, on the principle of the Bill, it is doubtful whether any amendment which would have the effect of undoing that principle, could be carried on a free vote.'

If the rebels lose tonight it is expected they will engage in talks with the Opposition in an attempt to agree a compromise which would have a chance of carrying the Commons at report stage.

One possibility is an amendment which would restrict the general application of the Bill to small shops, while drawing up a reviewed and enlarged list of other types of premises, like garden centres and newsagents, which would be allowed to open. An amendment on those lines won some backing in the Lords. . . .

Now compare it with the *Western Mail*'s version of the same story by their political man, Phil Murphy:

Up to 80 Tory rebels seem set to defy the Government tonight and vote against the Bill to lift Sunday trading restrictions.

But despite the rebellion the Bill is likely to be given its second reading by a small majority.

The Government compromise offer of a free vote during later stages of the Bill will probably be enough to lessen the impact of a revolt and the likelihood of a Government defeat.

This offer has been the only sign of any weakening of resolve from Mrs Thatcher and her Ministers. Tory MPs face a three-line whip in the lobbies – the toughest possible call for them to toe the party line.

Opponents of the Bill were still convinced last night they could muster the support of about 50 Tory MPs who would vote against a second reading.

Another 30 could abstain. If those figures were accurate they would take the Government close to defeat.

But Ministers believed the compromise offer had taken some of the sting out of any rebellion. Some sources were predicting the rebels might number a mere 30 Conservative MPs.

Ministers and Mrs Thatcher have made it clear there will be no further softening of their stance.

Home Office Minister Mr David Waddington said at the weekend, 'The opponents of the Bill have failed to come up with a compromise short of total deregulation.

'We will consider all suggestions but of those we have seen so far none commands any reasonable degree of support or has been workable.'

One of the leading Tories among the rebels, Mr Ivor Stanbrook, MP for Orpington, attacked the Government for its refusal to allow a free vote on the second reading.

'I am astonished the Government is not giving a free vote,' he said. 'If they cannot get this Bill, which is a matter of conscience, through the Commons on a free vote then they are not really in favour of choice at all. . . .'

The two papers are not very far apart in their structure and approach, or in their use of words.

Now look at the way this very different story of an odd human situation is presented in the *Daily Telegraph*:

Peter Sands had the shock of his life when he met a sister he did not know he had. Mr Sands, 51, from Eastbourne, had been working in the same office as his sister for five months without realizing they were related.

But Mrs Margaret Deadman was curious about her colleague in the vetting department of the Dental Estimates Board because she thought his name was familiar. 'One day she came up and started asking me about my family. When she said "I think I'm your sister" I got the shock of my life,' said Mr Sands yesterday.

Because of family difficulties, Margaret, who is 58, was brought up by her grandparents, five miles away from where Peter was living with his widowed mother.

. . . and compare it with this one from a London suburban weekly, the *South London Press*:

Daredevil bridegroom Dennis Todd had a windy ordeal getting to the church on time.

He planned to drop in on sweetheart Ingrid King – from 3500 feet.

But his parachute was blown half-a-mile from All Saints' Church, Blackheath, and he was forced to take a cab back to complete his grand entrance.

Ingrid (28) was waiting at the church, along with the vicar, the Rev. Henry Burgin.

The Rev. Burgin said, 'Dennis was only a few minutes late, but he was blown way off course.'

'We were more worried about a friend who jumped with him, as his main chute didn't open.'

But as guests looked to the heavens, Keith Kempton's reserve chute unfolded, and he landed safely, with Dennis and best man Micky Doyle.

Though the news focus of these two papers is quite different, the subeditor's approach, the fact sequence, and even the words and sentence structure in the two stories are similar.

If we turn, however, from general news to stories about the arts or the sciences, in which *The Times* specializes more than, for instance, the *Western Mail* or the *South London Press*, we find at once, especially with by-line writers, a more distinctive and technical use of words. These stories are aimed at people to whom the special vocabulary will be familiar. Here is part of a report in *The Times* on the proceedings of the British Association:

> The processes he described combined advances in the traditional fermentation of vaccine extracts with discoveries for manipulating micro-organisms by genetic technology to synthesize specific bio-chemicals.
>
> The substances which Dr. Butler, senior lecturer in bio-chemistry at Manchester Polytechnic, believed would become more widely available included the interferon family of anti-viral agents, and compounds called plasminogen activators which will dissolve blood clots.
>
> The plasminogens are potentially important for treating arthero-sclerosis, one of the main cardio-vascular diseases, although there is no established method for obtaining large supplies.

The difference between this style of writing and any story in a local weekly lies not in the method by which the essential facts are identified and projected but in the writer's use of a special vocabulary and an assumption that it will be understood by those who read the story, who might be a small section of the particular newspaper's readership.

Yet it is not just in *The Times* or in the even more specialized *Financial Times* that special vocabularies, with which the subeditor has to grapple, are found. Newspapers up and down the country have areas of news coverage aimed at particular segments of their readership in which special vocabularies appear. Take these examples:

Northern champion Clare-Maria Hall of Newcastle, said: 'The course is

400 yards too long. Into the wind, it's impossible to get up in two wood shots at some par four holes.'

The outward half with a par of 39 is particularly tough and the handful of players who returned low scores made their rounds on the homeward nine, where birdies were available down wind.

Only 20 of the 100-strong field broke 80 on the opening day and the cut to reduce the field to 32 for tomorrow's final two rounds was likely to be as high as 163.

The supply of consumer credit is likely to be increased by the reduced demands of the corporate sector given a slowdown in capital spending and the fact that profit growth on the back of lower oil prices will provide strong cash flows to provide a cheaper source of finance for investment programmes. In the public sector Government pressure for funding is unlikely to be onerous.

The Building Society freedom factor making some £5.5 billion of unsecured and similar secured lending and its finance possible from early next year is a major influence in credit supply even though the societies will at first tread warily.

The first example is part of a women's golf tournament reported in the *Liverpool Echo*; the second a report about borrowing money taken from the *Yorkshire Post*'s business page, but common to both is a particular readership assumption which is based on vocabulary, and on the use of accepted terms. The word use is special in that it is not aimed at the general reader.

Sport

The sports pages contain not one special vocabulary but a multitude to fit the many sports covered, some of them extremely esoteric in appeal. Take this example:

Both players used gentle services with impeccable control and exploited to perfection the boomerang which runs down off the penthouse roof flush to the back wall. Some exchanges were limited to three strokes: a boomerang serve, a weak return just over the net on to the roof and a volley kill shot under the tambour.

Perhaps only the enthusiast would guess that this is a report about a game of royal (or 'real') tennis.

With sport we come not only to special vocabularies and assumptions by sports writers but to a special 'voice' that sets sport apart from the news pages and takes it closer to the techniques found on the features pages, with which the remaining chapters of this book will be dealing. For a sports match report, though it might be classified as reporting, has more in common with the subjective approach of theatre or book reviewers in the 'voice' by which it addresses its readers.

Such reports, frequently under by-lines, utilize adjectives and emotive phrasing in their analysis of the style and method of teams and performers. Achievements and reactions are commented upon and compared, and results summed up. Take this example from the *Evening Argus*, of Brighton:

> Sussex snatched a breathtaking two-wickets victory over Leicestershire with only three balls to spare in the final championship day at Hove yesterday.
>
> With one game to go, against Glamorgan at Cardiff starting on Saturday, Sussex are now sixth in the table – the position they finished on last season.
>
> The Sussex innings, so full of good things among the mishaps, was dominated by Ian Gould whose 101 was his first championship century for Sussex and only the second of his career. His maiden 100 was for Middlesex at Worcester seven years ago.
>
> Two successive triumphs must put Sussex in good heart. . . .

Sports reports are intended for fans who want more than a kick-by-kick account and do not care if objective reportage is suffused with subjective comment and qualitative judgements. In local paper reports, partisanship in favour of the home team is expected of sports writers.

These points of attitude have to be borne in mind by sports subeditors. Accuracy of name and reference is still vital and the rules of language still apply, but there is greater emphasis on descriptive colour. A tennis report in *The Times* can be laced with words like this:

> The match had run away like sand through poor Mrs Lloyd's fingers. Miss Jordan has spread her game like opening an old chest full of spices, yellowed love letters and summer dresses. . . .

With 'name' sports writers this is what the reader expects to get.

Stories have still to be cast off to fit, however (sometimes very tightly in 'live' Saturday evening sports editions). The time factor on national Sunday paper sports pages is also tight with edition 'slip' pages going to press throughout Saturday evening, with constant rejigging of late scores and evening events. Copy on big boxing matches in the US might land on the sports chief-subeditor's desk in London at midnight, with 20 minutes left to last press time.

As one of the most specialized parts of the newspaper with its own readership and special vocabulary, the sports section has autonomy of production under the sports editor, who organizes both the reporting and features inputs (often contributing a column of gossip him or herself) and controls the editing through a sports chief-subeditor. Sports features are closely tied to current fixtures.

Sports pages, while slotting in to the overall production schedule, have their own pattern of deadlines (later than most news pages), and sports edition changes are more frequent than news in order to give edition areas a good coverage of local events and even local angling of national events. Some national papers using page facsimile transmission to satellite printing plant for the main news and features pages leave some sports pages open for local typesetting to accommodate late edition area reports.

Sports subeditors are themselves specialists because of the need for a knowledge of the rules and vocabularies of the various games, be they soccer, rugby and racing, or hockey, lacrosse or royal tennis.

On the pages, the methods of layout, use of pictures and headline typography still apply, but sports headlines make greater use of active verbs, names, and sometimes a play on names. These, again, are designed for readers familiar with names and terminology.

> REAL GLOS FINISH (about a county rugby game)
>
> QUEASY BUT EASY! LLOYD SERVES UP RIGHT MEDICINE (about a sick tennis player)

Financial

Here is another area of general newspapers where a special vocabulary is found – as can be seen in the example given earlier. Only specialist readers would read and understand the following:

> Traded options business reflected the preoccupations of the main stock market. As interest rate hopes look like pushing market indices back towards best levels and putting some glitter back on gilts, there were only two options in which investors showed significant interest.
>
> The stock market index contract registered 1667 trades out of the day's total of 8301, while the short-dated gilts option reached 962 contracts traded.

As with sport, this type of content requires from the subeditor a special knowledge of the references and vocabulary. Business news is also similar to sport in the amount of interpretive material that is used, with comparisons and forecasts occurring frequently in reports of company activities and currency, commodity and share dealings. Though less strident than in sport, there is a definable 'voice' in business news which proclaims its exclusiveness and reflects the writers' assumptions about their closed readership.

In terms of production, the business section is less demanding. Layout and display are usually straightforward and edition changes fewer. Business news benefits from the use of the computer in the constant updating of share prices and currency rates through link-ups with national and international financial services. Headline writing is most functional with a strong use of names and 'in' references. Examples:

ALLIED-LYON SHARES GAIN AS
AUSTRALIANS BUILD 3.84% STAKE
and
RACAL WINS
£18m ORDER

THE SOFT SELL

One area of the news pages where a clearly defined 'voice' can be heard is in 'soft sell stories'. This comes not from the use of a special vocabulary but through the personalizing of human interest situations through the eyes of their leading character (which can be an animal) in order to involve the reader. The method is used for stories about the oddity of human (or animal) behaviour or those in which the interest lies in the incidental details rather than in the main facts.

Take this example, from the *Daily Mail*, about two people who had a bizarre idea:

> Lovers Mike McCarthy and Amanda Tucker fell for each other in a big way.
>
> Once off the Eiffel Tower . . . and yesterday 328 ft from the roof of the London Hilton.
>
> The couple made their surprise leap after booking into the hotel under assumed names with parachutes hidden in their suitcases.
>
> Then at precisely 7.30 a.m. Amanda hurled herself into space above busy Park Lane followed immediately, above, by Mike.
>
> Twelve seconds later astonished commuters saw them land on the grass on the central reservation.
>
> Mike, 25, a London landscape gardener, said: 'This was no loony stunt dreamed up overnight. We'd both thought of doing it years ago and had planned it in detail.'
>
> They met at a parachuting rally in Holland and at Easter last year risked their necks on their first daredevil jump off the Eiffel Tower.
>
> 'It's a great hobby of ours,' said Mike.
>
> Their celebration was brief – breakfast at the Hilton – then 24-year-old Amanda, from Aberporth, West Wales, flew back to Florida,

where she is a commercial light-aircraft pilot. Mike will join her shortly.

The hotel, however, was very down to earth about the jump. 'Nobody here saw them,' said a spokesman.

The salient points about this story are:

1. The usual subediting approach of identifying and projecting the vital news point first is set aside and a 'delayed drop' intro is used (see Chapter 5).
2. There is a 'wait for it' story sequence in which the necessary explanation comes two-thirds of the way through.
3. The main characters are personalized on first name terms right from the first line.
4. The story is rounded off by a gentle pay-off line.

There is also a use of double entendre ('fell for each other in a big way') and controlled exaggeration ('hurled herself into space').

Thus what could have been a straightforward account of a rather dangerous stunt becomes a story in which the reader is anxiously involved in the strange behaviour of the young lovers. In this case the actual projection on the page included a picture of the couple safely together again and the headline LOVERS' LEAP. Such a story can liven up a dull day's news.

It often falls to the subeditor to present in this way a story that might have arrived in the office as straightforward narrative. The spur to this can be the picture to go with it, or maybe an idea of the page executive that here is a story that will benefit from the 'soft sell' treatment. Some subeditors specialize in this sort of story.

Here is a neat little personalized item from *The Sun* about a day in the life of a plasterer, presented again with a delayed drop intro:

Plasterer Graham Stent called round to repair 22-year-old Pam Jones's walls . . . and ended up delivering her baby.

Pam, of Cymmer, West Glamorgan, went into labour unexpectedly, and Graham, 31, rushed to the rescue.

He delivered a 5lb 9oz baby with the help of Ceri Robinson, who was visiting his girlfriend next door.

Note the use of the matey 'Graham and Pam' terms and the throwaway line about the man who happened to be calling next door.

Animals are favourite subjects for the soft sell. 'Nessie, the Loch Ness monster, may not only exist but has a boyfriend, scientists claimed yesterday,' begins a personalized story of serious scientific endeavour. Another delayed drop example begins: 'Tabby the tom cat is lapping up life again – with the hard-up master who gave him away. Intrepid Tabby made a 12-mile walk from his new home in Torrington, North Devon, back to 58-year-old Ivan Lee, of Barnstaple.'

The soft sell technique is not averse to using puns to point the oddity of a situation:

> A French bank robber used his loaf. He hid a rifle inside it and grabbed £2,000 from a branch of Credit Agricole in Lyon.
>
> The bread may not have been fresh, but it was deadly. And a bank teller had no hesitation in handing over the dough.
>
> The robber, about 25 years old, wearing jeans and a khaki jacket, had carefully carved out the interior of the baguette to conceal the 22-calibre rifle inside.
>
> He walked out of the bank and disappeared in the traffic.
>
> Now police are searching for any crumb of evidence they can find.

The subeditor in this *London Evening Standard* story wisely sticks to only three puns. More could have ruined the story.

The soft sell treatment requires great confidence in the subeditor that the titillating morsel he or she is offering the reader can be adequately backed up. The intro must contain evocative or tantalizing words and not be just a variation of 'once upon a time'. It is like lighting a slow fuse that presages the big bang. The spark must be there or the story will be a damp squib.

When we read the 'lovers Mike McCarthy and Amanda Tucker fell for each other in a big way', we know something is afoot. With the next sentence we are baited: 'Once off the Eiffel Tower . . . and yesterday 328 ft from the roof of the London Hilton'.

Likewise with 'The French bank robber used his loaf'. Next move: 'He hid a rifle inside it'. Even if the whole story is not ready to be revealed at once, the subeditor must whet the appetite of the reader so that he or she becomes captive.

The advantage with the soft sell treatment is that it makes direct contact with the reader to whom it appeals with colloquialisms, intimate asides and matey first name terms. In its turn it relies on its acceptance by the reader as a special voice on the news pages that proclaims this story as being different from the others.

Facetiousness

Subbing a soft-sell story takes time. The material has to be carefully balanced, and every phrase lovingly polished so that the reader totally accepts the artificiality of it. The fact that the normal intro-sequence conventions do not apply makes each story a custom-built job. Yet the approach is fraught with danger in the hands of a subeditor who pushes puns and humour over the brink into facetiousness or who tries to deal humorously with something that is not intrinsically funny.

Keith Waterhouse, in his book *Daily Mirror Style*, warns against facetiousness which he defines as 'a form of sustained banter. It joshes the reader from sentence to sentence, rather like an old-fashioned club comedian . . . interpolating such expressions as: 'Ere lady!', 'It's true!', 'Listen, missus', 'He did, honest!' and so on between the paragraphs. And he warns, 'Automatic punning is a tedious schoolboy game'.

In other words, a news page on which subeditors have been allowed to jolly up the stories with puns and facetious headlines and a rash of delayed drop intros will finish up looking silly and juvenile and lose all authority.

IS THERE A TABLOID STYLE?

The changes in presentation that came about with the ideas first tried out in the tabloid *Daily Mirror* from the late 1930s onwards used to prompt the question: Is there a tabloid style? The use of bold display, heavy sans headlines and short crisply edited items, together with comic strips and eye-catching pictures, made a break with the old broadsheet dailies and set up a new popular market. Today, when several middle-of-the-road papers have become tabloid shaped and the *Daily Mirror*'s revolutionary ideas have been either copied and absorbed by other papers, or modified and made uncontentious, it is less certain that there is a fundamental difference in style between broadsheet papers and half-size tabloid ones.

If, for style, we substitute voice, however, there is some merit for claiming for *The Sun*, regarded by many as a successor to the old *Daily Mirror*, a distinctive place in the tabloid field, not so much for its display, which uses the now well tried tabloid ideas, but for its words. And not just its words but specifically its adjectives.

Adjectives, frequently excised in newspaper practice in the crusade for tight subbing, have come to form a deliberate part of the word pattern with which *The Sun* evokes for its readers a glossy away-from-it-all world full of megastars, superdads, luscious Lindas and endless jollies. In what other paper would the reader encounter, in one issue, the following adjectives occurring in headlines, intros and captions?

staggering (twice)	stunning
luscious	terrific
super	dashing (twice)
sizzling	sexy (three times)
amazing	curvaceous
beautiful (twice)	scorching

And, to test the consistency of the emotive word pattern, another selection from an issue six weeks later:

tragic (twice)	gorgeous
gigantic	beautiful (twice)
sexy	pretty
lovely	legendary
super (twice)	staggering
amazing (twice)	lucky

The choice of adjectives is not an accident. It is the clearest evidence of a deliberately orchestrated voice by which *The Sun* newspaper addresses its readers and seeks to put excitement into their lives.

The approach is built around short, bold, fact-laden stories (as many as six or seven to a small, well-illustrated tabloid page) with evocatively worded headlines in the same issue such as:

> SIR, GET YOUR
> HANDS OFF MY
> FLAMING
> BOTTOM

or

> WHY I STRIPPED OFF

or

> PAL STOOD IN AT JAIL AS
> RANDY CONVICT ESCAPED

In its editorial opinion, *The Sun* reveals the secret of its special voice – mimicry. Its editorial opinion conveys in no uncertain way the views its readers might be expected to voice in the local pub, and in the sort of language they might be expected to use:

KEEP 'EM IN
Home Secretary Douglas Hurd's plan to beat the prison officers' dispute – the mass release of crooks – is barmy.
   True, the lucky lags will be non-violent offenders with less than six months to do.
   But most of them will already have had hefty remission so why should they get any more?
   Instead Mr Hurd should send in the army to bang up the lot until the dispute is over.

And that's telling 'em!

JOURNALESE

There is one special language – and vocabulary – that newspapers would do well to be rid of, and that is journalese. This is strictly a debasement of the language caused by the infiltration into the text of the stereotyped short words and 'short cut' phraseology used in headline writing.

Words that are tolerable in a headline because of the difficulties of character count and the need for visual balance should be examined closely before they are used in the text. Here the need is for precision. The explanation and justification of facts in the text, upon which the headline leans, cannot be put at risk by ambiguous jargon words, however well these might have served the purpose in drawing the story to the reader's attention.

While the short word 'boss' might have got the subeditor out of a difficulty in the headline, it must be clarified and taken a stage further in the text as manager, supervisor, chairperson, general secretary or what-ever. 'Boss' is simply verbal shorthand.

The words 'rap' or 'row' in the headline imply criticism or disagreement, but they do not have the precision that the text requires if the story is to mean anything to the reader.

To be 'axed' or 'probed' in a headline might put across the general message (if there is no time to write a better headline) but the text has to back up these words with clearer, more precise ones.

Words of this sort are used in headlines not just because they are short but because they can fit a variety of stories in which more specific descriptions are too long to be accommodated. It is this lack of precision (apart from the fact that they have become headline jargon) which limits their usefulness in the text. The fault with journalese, in fact, is not just its mind-lulling jargon but its vagueness.

We have noted elsewhere that words used in news language are rooted in everyday demotic speech, even though the grammatical structure and thought and fact sequence are not. It is a useful test of a piece of edited news if the subeditor mouths it silently to judge by its 'sound' if the words used are real demotic words or chunks of a newspaper's home-produced headline jargon.

Would a neighbour say to you: 'I hear top cops have quizzed Jones following a cash probe at the superstore he's just quit.'? And would you reply: 'Yes, and now he's quit home in a night drama after a big rap from his wife.'?

14 FEATURES: PLANNING AND DESIGN

Features is broadly that part of the editorial content containing opinion, assessment and advice – the subjective rather than the objective material found on the news pages. In practice, the features department of a newspaper deals with all manner of non-news content ranging from crossword puzzles to articles by the famous, and taking in such things as women's pages, television gossip and agony columns.

In production terms, features work calls for differences in routine from the news pages. There are a number of reasons for this.

The features department is to some extent the counterpart of the news room, with the features editor in charge of writers and copy inputs, but the work is more programmatic, with some autonomy being allowed to specialist writers to provide up-to-date features on their topics. These tend to be used on a regular basis on pre-selected parts of the paper, such as the political or leader pages, the women's pages, or the showbiz section. The quality Sunday papers even have separate magazine sections given over to a variety of features.

There is more forward planning with features and more time spent on their preparation. In the case of serialization of important books, for instance, there can be weeks of negotiation and discussion followed by a careful 'gutting' of the book to provide a series of instalments. Showbusiness writing, fashion articles and holiday supplements are other areas of features where the writing is tied to events and is planned well in advance.

There are always background features to be provided on contemporary topics as they arise but, on the whole, the features editor is able to keep a much longer term work schedule than the news editor and the department's workload can be programmed without the unpredictable flow of copy that happens with news inputs. Since many features require in-depth investigation this longer-term planning is a necessary characteristic of the work of the department.

COPY SOURCES

The longer time-scale, guaranteed columnage and the need for persuasive, opinion-forming writing means that there is more scope for creative prose, or writing on which writers can impress their own personal style. Features departments usually have a few gifted stylists who can turn their hand to a variety of contemporary topics under by-lines which become, in effect, brand names. These are the general 'name' feature writers, whose role is to provide comment on, and back-up to, the news.

Many feature writers, however, specialize tightly, using their own particular knowledge, experience or training on subjects such as politics, education, economics, technology, fashion, motoring, travel and even hobbies. Some of these are staff writers, where the demand for their subject is regular. Some are outside contributors who are paid a retainer or a fee per article and who might write for other publications on their speciality, not necessarily under their own name. Some newspapers share the work of group correspondents, working under the same ownership, for this purpose.

Features material such as crossword puzzles, comic strips, cartoons and horoscopes are syndicated through agencies who specialize in these things and operate on a national or even international basis. Some syndication agencies also provide topical features and world-wide background situationers.

Certain special articles might be ordered from non-journalist experts such as ex-Cabinet Ministers, sports stars, trade union leaders or local council leaders who happen to be in the news or are experts in their field. Equally, freelance writers who submit material 'on spec' might find their work chosen if it fits a newspaper's plans.

Readers' letters form a significant part of a newspaper's features content, not only as general letters pages, but as the basis of advice or service columns written by experts ranging from travel writers and financial and legal experts to the 'agony aunts' who deal with readers' emotional and personal problems. Some staff experts give advice by letter as part of a readers' service as well as basing their column on letters. Letters might also be solicited on specific subjects where reader-participation features are intended (see F. W. Hodgson, *Modern Newspaper Practice* (Heinemann, 1984) pp. 57–71).

Copy sources are thus more diverse than with news, but at the same time more pre-planned and subject to precise arrangements (see Figure 39).

PLANNING

Filling the features pages forms part of the same planning procedure, by

228 *Modern Newspaper Editing and Production*

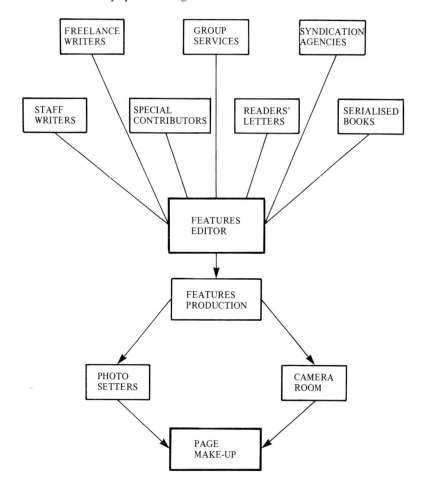

Figure 39 *The flow of work through the features department of a big modern newspaper*

discussion and conference, as the news pages. A good deal of the material is tied to the news. At executive level ideas are discussed and exchanged and spin-off features initiated for the day or for subsequent days.

Usually, however, there is greater participation by the editor, since many features express the opinions and policy of the paper for which he or she is responsible, though some writers have a free hand, within certain limits, to express their own. There is a close relationship between the political and industrial writers in the planning of the daily editorial opinion, or leader, which is the official mouthpiece of the paper. This can be written by the editor – though bigger papers employ one or more special leader writers – and it is another subject for conference discussion.

Where there is particular topicality to a writer's subject, such as a controversial television programme or the launching of a new car, or a worrying phase of industrial trouble, special space might have to be allocated. Sometimes writers are asked to provide a 'comment' piece to give added depth to a news story. Running a piece by a big-name outside contributor can involve providing a page one blurb on the day or the day before so that the reader knows on which page to find it. Contents bills might be ordered to advertise an important feature, or special placing given to news stories that are connected with the subject. All these matters are brought up at the planning conference.

Inevitably there is some overlap with news since by-line experts might be called in to report stories relevant to their field as well as provide features. In the same way, reporters might be asked to help out with investigative features where leg-work is needed. In some newspapers, where by-line writers are allowed to inject comment into the news they are covering, it can be difficult to tell whether a piece is a news item or a feature. This is a controversial area of newspapers where editors who allow this overlap might encounter complaints about bias in news coverage.

Topicality remains a strong element with features as it does with news. If subjects do not arise out of the current news then they are pegged to seasonal or contemporary events, as with show-business and fashion writing and travel, arts and family features. Even writing on food and recipes tends to have a seasonal bias. Features thus relate to the news pages in the sense of timeliness, and the volume and content of each has to be established at the outset to the editor's satisfaction so that the two parts are complementary.

The difference, in production terms, is that, with the exception of one or two 'live' pages, features pages are planned, typeset and made up in advance of news pages on all newspapers, some pages in morning and evening papers being prepared the day before.

The reasons for this are twofold:

1 Apart from features which are spin-offs of the news of the day, the longer-term ordering and planning in the features departments ensures that features copy and pictures can be made ready earlier and often to a pre-determined length.
2 Since many features are prepared for fixed slots or to be displayed on pages on which the space is planned and guaranteed (often with the earliest sold advertisements) and are not affected by the ebb and flow of news situations and pictures, the layouts can be got ready without hindrance or delay.

The consequence of this is that editorial production is staged out so that subediting and typesetting of late news does not find itself clashing with the setting of the television programmes, articles on the home and last night's theatre reviews. Likewise, features page make-up, whether paste-up, on screen or in hot metal, can be completed first, and space, staff and machines made available for the later pages.

On morning and evening papers where production splits into shifts, the earlier shift deals with the features pages. On some papers, features production is separated from news and there is a features chief subeditor, responsible to the features editor or the assistant editor (features), where there is one, with separate subeditors. In some cases there is even a separate art desk for page design.

The features page that is left until last is the one containing the editorial opinion column (which needs to be up to the minute) and the main feature article of the day, and perhaps the political column. This page usually has an editorial production slot not far in advance of the last news page so that it can reflect, and comment on if need be, any important late news situation that crops up.

CONTROL

The control of features production varies, depending on whether or not it is a separate operation from news. The function attaching to titles also varies from office to office. In some cases the features editor wears a features production 'hat' as well as a features gathering 'hat' and supervises the make-up of the pages. In bigger offices there is sometimes an assistant editor (features) who is the controlling production executive, with the features editor's function being more like that of the news editor.

On most small papers the chief subeditor supervises the production of both news and features pages, but works closely with the features editor who might scheme the main features pages himself or herself, leaving the chief subeditor to deal with things like television programmes and features in regular slots. The copy, with instructions, is passed out to the subeditors' table, though for special stories even the smallest newspaper usually has one or two features subeditors who work either directly under the features editor or on the general subs' table, and to whom the work is given.

On bigger papers, and certainly on national dailies, there is generally a division of labour, with the features chief subeditor supervising a separate subbing operation under the direction of the features editor or the assistant editor in charge. Under this system the features back bench initiates its own page layouts which are then passed to the art bench for detailed drawing. Where a newspaper is separated strictly into news and features pages this offers no problem. Where, as sometimes happens, there are some hybrid news/features pages the editing of the features is carried out by the features department while page make-up is supervised as part of the main editorial production.

It will be seen that there is some variation in method in the editorial production of features pages depending on the manning levels of each paper. There are also differences in the approach to editing and layout which will be dealt with in the rest of this chapter and in the following chapter.

PAGE DESIGN

Instead of a balanced arrangement of news stories with a lead, a half lead and maybe six to ten other stories formed into a pattern based on headlines, pictures and text, features pages have maybe two or three items, and sometimes only one. As a result, the projection is more contrived and thematic than that of news, with more emotive or eye-catching headlines and a deliberate visual pattern. Typography and pictures are used with greater freedom.

Typography

To offset longer texts the typography is generally bolder and more varied, with quotes and subsidiary statements used as breakers. Panel rules often divide a feature from the remainder of the page, and stars, blobs and reverse type (i.e., white on black or black on tint) are sometimes used to highlight parts of the text.

The type chosen for the main feature headlines is likely (though not always) to be different from the standard news type, and the choice of type character can be influenced by the theme of the text. Thus, depending on the type character of the paper, lighter faced type such as Spartan, Record Gothic Light, Futura, Century Light and the old light seriffed faces such as Garamond might be used for softer features, while bold expanded faces are brought out for a feature headline that needs impact.

Type decoration can range from the holly borders on Christmas features to the crowns and 'architectural' rules and borders kept for royal features, with drawn diagrams, motifs and bleach-outs sometimes being used to supplement half-tone illustrations, although artistic 'white space' can be an important element.

Pictures

Pictures on features pages are often bigger or are likely to be used in series or as part of a compo or montage, and are generally more integral to the display. Where a picture is usually chosen for a news page for its news value, on a features page it can be used to establish mood or atmosphere, or can have a more deliberate display function.

The text

The text usually has the same body type as the news pages but the measure, or setting width, tends to vary from the set columns. In fact, features page setting is often deliberately of different measure from the news pages.

Drop letter capitals on intros, now little used in news pages, are sometimes house style on features to decorate the text. Indented setting is commonly used to show greater white space, and column rules, which are style on news pages, dispensed with and columnar white space used for separation instead.

Crossheads are sometimes chosen from display faces to match the headline type rather than conforming to a general crosshead style, and can be of two lines instead of one, sometimes with underscores. Italic or bold paragraphs to give emphasis, or as eye breaks, are more likely to be found on features pages because of their usefulness in breaking up long texts.

Other devices are the use of *stand-firsts*, which are short 'blurbs' in different type placed above the intro (i.e. they stand first) to introduce or explain the feature or its author; and the use of writers' *by-lines* and *sign-offs* in display type.

It would be wrong, of course, to use all the typographical devices mentioned above on one features page, or in one newspaper. What we have noted are the sort of devices that are used. Even though there is a freer approach to features page layout, newspapers usually have a style, either written down or accepted, encompassing the sort of types and devices that may be used with features text so that the pages do not become a typographical mish-mash.

GUIDING THE READER

Stock *logos*, or identification devices (Greek λογος: word), are commonly used to flag regular features such as television programmes, cartoons, horoscopes and the editorial opinion column. Where possible these features editorial appear in a fixed spot in each issue of the paper to facilitate easy finding. The logo message is simple: *Opinion, Your stars, Sunday fun*, or simply the writer's name in the case of a regular name columnist who comments on current events. The name together with picture becomes the brand image by which the columnist is displayed, the headline being of less importance. In some cases there is no headline at all. The reader knows what the name conveys and the sort of style to expect.

The 'selling' here is low-key and such stock features might have their regular slots on pages which are otherwise given over to news, but are differentiated by their logo and static familiar display so that there is no danger of mistaking them for news.

The typesetting and flagging of TV and radio programmes is an important area for the familiarity technique, and newspapers take a great deal of trouble to give programmes a format which makes the details readable and instantly findable as well as accurate. Bold type is used to highlight the name and times of programmes and 'still' shots from programmes are included to make the page attractive.

These contrasting elements of bold display on the one hand and static familiarity on the other can be seen in the *Liverpool Echo* features spread reproduced in this chapter (Figure 40). It shows an imaginative design for a main feature, built around pictures, while on either side of the spread are static panels enclosing the editorial opinion and the paper's name columnist, each topped by a recognition logo.

Series layouts, especially in Sunday papers, require both a deliberate well pictured display to catch the eye and carry the 'big read', and a familiar logo to mark each instalment and to remind the reader about it.

It will be seen from the page examples analyzed below that headlines can be subservient to pictures in drawing the eye to the page material. In some cases the headline is relegated to the position of a decorative breaker in the text. It will also be seen that there is a good deal of variation in the way in which type and pictures are used, depending on the readership market of the paper.

By-lines have more importance on features than they have on news. Where a news story might carry, say, an *Evening News* reporter line or the reporter's own name in small type, or even no by-line at all, a feature by-line is often given typographical prominence because the writer's identity and opinions are relevant to the text.

The by-line is usually set in a display type, perhaps inside panel rules, and is used either above the intro or as a breaker towards the top of the text area, or as a bold *sign-off*. The name might be accompanied by the credentials of the writer, particularly if he or she is an important outside contributor or special expert. In the case of some very exclusive stories by famous people, the writer's name can be in the largest type on the page.

FEATURES WORKSHOP

The pages reproduced in this chapter are intended to show the freer

Figure 40 *How a big town evening paper, the* Liverpool Echo, *handles a double-page features spread, using photoset materials*

approach to features page design, whatever the type of newspaper, and to demonstrate the ideas that have been discussed above. They should be compared with the more static design effects in the examples of news pages in earlier chapters.

The two-page spread from the tabloid *Liverpool Echo* (Figure 40) is a good example of the imaginative use of type, half-tone and display devices in presenting three ingredients across the pages' normal fourteen-column width. The absence of advertisements has enabled the designer to split the spread into three panelled segments of equal depth but of varying width. Nearly two-thirds of the space is devoted to the main feature which consists of two connected stories about ways in which people in a big city are coping with the despair of unemployment.

The picture content is all in the main panel but its spread across the gutter enables the pictures, a horizontal and two nicely balanced verticals, all black-edged, to dominate both pages, and the panel on the right manages, without any ill effect, to survive on a small head shot with the by-line. The deep vertical of the boy boxer in the centre, an excellent picture cropped a little close at the bottom, is placed skilfully so as to separate the two main type masses, while lining up at the bottom with the horizontal picture on the left-hand page to enable a secondary headline to cross the gutter and thus help to hold the main panel together.

The headline type of the big panel is in the stock sans headline type of the paper (which has a predominantly lower case sans format) but it gains in vividness by being used in a large size in short lines climbing down the page, and by being given a 4 pt underscore and extra white between the lines. The use of a bold white-on-black strap line cross the top of the pages, and winged into the heavy 8 pt black rule enclosing the panel, knits the feature together and is an instant eye-catcher. It leaves the reader in no doubt as to where attention is being directed.

The bottom picture, another good one, is used to separate the text of the second part of the panel from the first so that a long read is well broken. Other useful devices that mark the spread as a features display are the white-on-black stand-first and the under-scored crossheads, although the first column could have done with a second to give balance. The second item in the panel is also set in a narrower measure in 8 pt bold.

The panel on the right is rightly presented in a softer key. In fact, competitive type and pictures would have ruined the spread and taken attention from the main feature. It is a by-line column of miscellaneous items under a name known to the readers since it simply carries the logo

Walter Huntley. Subsidiary headlines in Ludlow Black (a thick serif type) and a condensed sans face in varying sizes keep the eye busily moving round the text. The body type uses light and bold in five different measures (a little extreme!) including a 'panel within a panel', with smaller nuggets being divided from the rest by 8 pt star rules. It is altogether a useful foil to the long read of the main feature.

The main headline is an expanded version of the paper's stock type, falling rather short on the right, and the whole is enclosed in 2 pt panel rules. Not to have had these rules would have weakened it visually against the big panel.

The third item is the editorial Comment panel which is set across 11 ems in 2 pt rules in its regular position on the left-hand side of the features spread, and in 10 pt of a light sans body face not used elsewhere. It is modestly garbed in a two-line 30 pt headline, small 'Comment' white-on-black, and is without crossheads. Adjoining the first long column run of the main feature this produces the only visual flat spot on the page. A clever stock cartoon, 'Kidstuff' under the Comment panel, a curious placing, helps to make up for this.

The similarities to the *Liverpool Echo* spread are more immediately apparent than are the differences in the arts page of *The Guardian* which is derived from a paste-up based on photoset material (Figure 41). There is the same use of heavy rules and juxtaposition of bold vertical and horizontal black-edged pictures and the use of standfirsts to introduce the items. Yet with this *Guardian* features page we encounter the contemporary features style that has been developed by the quality national papers. Here the design highlights the reading text rather than the headline, which is relegated to a terse label subserving the pictures.

The downgrading of the headline in this design philosophy is apparent in the fact that on this large seven-column broadsheet page with modest advertising there is no headline wider than two columns. The main feature, in fact, is introduced by a four-column 14 pt Ludlow lower case standfirst, and is 'sold' to the reader on a daringly cropped deep two-column half-tone picture of author Anthony Burgess, which has a simple 7 pt line caption, and divides the text into two long legs. By comparison, the headline, though in 48 pt Ludlow to match the standfirst, serves the purely decorative purpose of introducing white space into the middle of the page to help offset the length of the text. Its message, OPERATION OBERON, far from informing the reader, demands that the text be read first to invest it with meaning.

The other two main pictures are likewise daringly cropped and demonstrate how, in cropping, a picture of a head can survive losing a good deal of the forehead as long as the chin is left unscathed. In fact the virtue of this low cropping is to accentuate the eyes, which is the characteristic of all three of these pictures.

Again, with the two stories introduced by these pictures, the headline is a

Figure 41 *New design ideas at work in a features page from* The Guardian

bold label providing balancing white space, the informative message to the reader being conveyed in the standfirsts.

In all three stories the run of text is long and unrelieved by crossheads, a characteristic common to the quality papers whose words-conscious approach rests on straight text setting, eschewing such frills such as crossheads, reverse indent, bold paragraphs, blobs or drop letters. These are left to the popular papers on the assumption that quality paper readers do not need such help or visual diversions.

The review items below the fold have no overall headline at all but are held together simply by a piece of 6 pt rule crossing the four columns and cutting them off from the main feature above. This, with a pica of white space above and below, manages to do the job adequately. Each item is identified by a simple box giving theatre and reviewer and a one-line 18 pt lower case headline using only the name of the play.

The two remaining pictures, single-column black-edged head shots, form useful focal points and text breakers in columns two and four and help to offset the rigid plainness of the text setting. The only other element that takes the eye, other than the logo Arts Guardian on the top left, is the characteristic use of vertical 2 pt rules and horizontal 6 pt rules, both well whited, as dividers. This ploy, too, helps to make the plainness of the text acceptable to the eye, where leaving white space in a layout like this would have failed. The headline type, other than the seriffed logo, is a nice balance of a light sans lower case and the chunky Ludlow.

The overall effect of this *Guardian* page is to demonstrate how text can be made to look interesting by personalizing it to the reader through the use of cleverly chosen, well cropped pictures which dominate the eye and separate the items. It has good visual balance while at the same time making display subserve the aim of giving the reader plenty to read.

The leader page from the *Daily Mail* (Figure 42) shows how elegance and boldness can be obtained through the traditional use of pictures and typography allied to emotive features headlines. In terms of display the page is differentiated from the *Mail*'s news pages by the use of Rockwell slab serif type for the headlines and by-line on the main piece, as a foil to the paper's Century Schoolbook type format. The picture of the child in close relation to the 84 pt Rockwell Bold lc headline, STARVED INTO SUBMISSION!, forms a vivid focal point that takes the eye. The lighter sans headline alongside the intro adds explanation, which is encountered by the eye as it moves up to the smaller pictures at the top of the page. The word 'starved' links in the mind with the emotiveness of 'infamous'. The caption on Stalin and Mengistu confirms the fact that this is an angry piece about famine and military rule in Ethiopia written from the writer's heart. It is a judgement, an assessment based on facts *and* opinion, a passionate plea to the reader for agreement of view.

Figure 42 *A features page from the tabloid* Daily Mail *showing the use of slab serif type for the main display*

The layout demonstrates the visual power that can be achieved by letting the pictures carry the top of the page, while using the main headline as a pivot for the text to spill round. The dominance of the three pictures is assured by enclosing each in a 2 pt rule, and the unity of the whole stressed by the pica black rule separating the feature from the leader column. The leader, or editorial comment, is in its familiar left-hand position, and is set in reverse indent. The headlines on both leaders are in Medium Condensed Gothic lower case, the *Mail*'s regular recognition type for its daily

Figure 43 *Discreet use of type and borders break up the text and yet give a well filled look to this features page from a country weekly, the* Chichester Observer

Features: planning and design

Comment. The reading text on the page is broken up by crossheads and italic paragraphs in the main piece, while the third item, at the bottom of the page, is set in bold lower case with a sign-off instead of a by-line – a modest but effective use of traditional features ploys.

The page is an example of the movement of middle-of-the-road tabloid papers towards the wholly lower case format pioneered by the broadsheet quality papers, though the use of Century lower case news type on the third item is a pity in an otherwise pleasingly balanced design.

With the features page example from the *Chichester Observer* (Figure 43) we turn to the weekly county press, in this case a well-filled 52-page broadsheet paper which demonstrates how the use of a direct input computerised system need not alter the old world charm of a page of country topics. All lower case Caslon headlines of modest weight introduce the subjects with simple labels: VERY ODD TRACKS IN THE SNOW (nature trails) and THE DANGERS OF THE BIG CATCH (angling notes). Against fairly bold advertisements, the features editor has opted for a four-column line drawing of badger tracks and has created focal points by flagging each item with strip white-on-blacks which identify them instantly by giving the names of the writers, who are regular contributors to the paper. The main columnist, Geoffrey Godber, gets three legs of bastard setting at the top of the page, the other features being in roman single column across the ten-column page. The whole presentation gives an impression of a good read for a leisurely moment in a style that is only possible with the broadsheet format – the ideal medium for a county weekly.

15 FEATURES: LANGUAGE AND PROJECTION

The broad principles of features subediting are the same as those of news. Copy has to be checked for factual and grammatical accuracy, edited to the length required and freed from ambiguity, wordiness and errors of house style. Headlines and captions have to be provided, and the whole marked up either on copy or on screen, where the facility is used for subediting, with correct instructions so that it is delivered from the typesetter in the type, size and measure wanted for the page.

Page layouts, as with news, have to be visualized and drawn, the types chosen and the material projected in terms of headline, text and pictures so that the finished page corresponds with the ideas of the page executive who originated it.

Yet, having agreed the mechanics, we are aware of a difference of stress at various points in the editing and production routine. Some aspects become less exacting, others more so, and there are important differences of approach in the editing.

ACCURACY

First warning sign: it is dangerous to assume that experts always get their facts right. Some who are close to their subject rely too much on their knowledge and memory and let through basic mistakes where the less informed would have checked. Memory can play false with the initials of someone, or their title or function, or the year of a Parliamentary Act. Politicians, for example, are notorious for factual blind spots, as special contributors, when not being briefed by their researchers. It is always best to add up figures and check statistics and tables in stories to see that they work out. Dates are a frequent source of error.

On the whole it is with the one-off non-journalist contributor, unused to newspaper discipline, where the greatest likelihood of factual error lies.

Since such contributions are usually specially ordered opinion and comment pieces which will be well displayed and widely read, they throw a more responsible burden of checking on the subeditor. It is no satisfaction for the editor to find that an expert contributor got it wrong when subeditors are being paid to spot such mistakes.

With regular staff writers the danger is less but the responsibility remains. Advice columns, for instance, upon which readers rely, cannot afford to make mistakes. Merchandizing information in fashion pages has to be right. Crossword puzzles have to be checked to see that the clues work out. Television and radio programme details, a tedious editing chore that can fall to new features subeditors, are a potential minefield for the unwary.

Mistakes in figures in newspaper competitions – usually looked after by the features department – can result in bad publicity and cost newspapers money. An even greater danger arises from the failure to spot factual errors or unjustified statements that might get the paper into legal trouble (see Chapter 10).

The wider range of material on features pages and its often controversial and provocative basis, thus makes checking for accuracy a high priority in features subediting.

LANGUAGE

Because features are written mostly to a length by specialists and because they are often important for the way they say things as much as for what they say, they are not normally given 'heavy' editing. Rewriting, carried out on news stories usually because of bad copy, complex copy sources or shifts in emphasis, should not be needed. While there is a freer approach to intro, structure and sequence, these are normally imposed on a feature by the writer. A feature that needs re-nosing and re-structuring by the subeditor is the exception.

In the subjective writing found on feature pages, clarity is still the aim, but the text is free of the pressure on space and the relentless word economy of news subediting. Feature writers writing under their own by-lines are permitted a wider range of vocabulary and idiosyncracy of phrase. In some specialist areas appealing to a segment of the readership a special vocabulary might be used as, for example, in articles on space warfare, economics and some areas of sport.

Likewise, heavy cutting caused by demands on space from other stories, the revising for later editions, and the organizing of running stories, are normally absent from features subbing. In this sense the work lacks the excitement and 'dicing with the clock' of news subbing.

What has to be allowed for is a greater regard for style in the sense that the copy contains argument and description by which specialist writers communicate more directly and more personally with the readers than do news writers. This can involve longer sentences and longer paragraphs, though not excessively so. 'Repair' and improvements can still be necessary, but in correcting any weaknesses of grammar or sequence, the subeditor must be careful to preserve the writer's style. Style is a fugitive thing that can be damaged by insensitive or unnecessary editing and it is here that the greatest difference of editing approach lies.

Here are two extracts showing style at work in features. These should be compared with the examples of news writing in earlier chapters.

> Nor has any link been established between economic prosperity and the rapid replacement of old buildings. Paris, Geneva, Munich have prospered with blanket conservation policies more effective than London's. The demolition of central Bradford and much of Newcastle – once both magnificent towns – brought no particular wealth. The most conserved cities in Britain – Westminster, York, Bath – scenes of the fiercest postwar conservation battles, are now among the most prosperous.
>
> Tourism is Britain's fastest growing private industry, still woefully undercapitalized and often mismanaged. In many places including London, it represents real future wealth, and it depends on conservation.
>
> What is wrong with British conservation is not its existence or extent but the manner of its implementation. It is afflicted with the curse of British planning: over-attention to detail yet ineffectiveness in the grand design. We can save an architrave and ban a mews garage yet not prevent the London docks being left a bleak wilderness.

> Our Christmas feastings are nothing to those of our ancestors. From medieval times until Cromwell's Commonwealth, the Twelve Days of Christmas were celebrated with feasting that could reach wild excess. Then the Puritans abolished Christmas festivities. A popular song of the time ran:
> > Plum broth was Popish, and mince-pies,
> > Oh, that was for idolatry.
>
> With the restoration of the monarchy under Charles II, feasting returned, though never to the same extent, and favourite Christmas foods could be safely enjoyed once more.
>
> Plum broth, or Christmas Porridge or pottage, was the forerunner to the Christmas Pudding. A rich soup stuffed with dried fruit, thickened with breadcrumbs and enlivened with alcohol, it is quite delicious. Pepys noted on Christmas day 1662 that he enjoyed 'a mess of brave plum pottage' before his roast beef.

In the first piece, Simon Jenkins, a guest feature writer, is examining in the *Sunday Times* the effects of planning decisions on cities. Information and opinion are dovetailed together and the paragraphs consist more specifically of thought sequences than is usual with news. The material is held together by the swing of the argument and the style flourishes on the use of antithesis and rhetoric. 'What is wrong with . . . is not its . . . or its . . . but its . . . etc.'

The second, from the *Daily Telegraph*, is an example of the folksy approach that can work with old-fashioned subjects such as a feature on Christmas recipes. Titbits of information are enlivened by quotations and a cascade of mouth-watering verbs.

It is important to keep the colour, feel and pace of a well-written feature, and to regard editing as a means of preserving and enhancing these qualities – in addition to checking that the facts are right.

Some by-line contributors whose style is their brand image have contracts that specify that their copy should not be altered without their agreement. Editors try to limit the number of these arrangements since they can constrict production. Also, however good the writer, perfection does not exist, and there can still be space problems due to changes in advertising or writers exceeding the ordered length, or cases where parts of a feature are not liked for legal reasons or on grounds of taste. The duty to amend remains with the subeditor.

If one is looking for sacred cows then the one thing above all others that should not be cut or altered by a subeditor is the paper's editorial opinion piece, although the facts, dates, quotations, etc. that it contains still have to be checked. An error here that gets into print is akin to an act of sabotage. This is definitely an area where, if the copy over-runs, the editor will want to cut it.

Ghost writing

In some newspapers, notably the national Sundays, the practice of ghost-writing might be found. This, to the lay person, means a feature (or even a book) that is written by a professional writer for the person under whose name it appears.

Ghosting might at first sight seem an indefensible practice. It is certainly not one that newspaper journalists want to see encouraged in a general way and yet it can be the only means by which important one-off pieces of writing can reach the reader. It is used where a newspaper wants to carry a personal story of someone in the news or whose experiences and opinions are of great potential interest to the reader but who might be too busy or have insufficient skill with words, to provide copy in an acceptable form in time for the edition.

A feature writer who is an experienced 'ghost' is put in with the person by agreement, often with a tape recorder but sometimes with a notebook, and 'talks out' the material. The feature is then written, keeping intact the opinions and verbal style of the person – who might, perhaps, be a trade union leader, a sports personality or even a politician.

When complete, it is shown for approval to the celebrity whose name it will carry. Any necessary adjustments to wording or content are then made so that the finished article is agreeable both to the 'author' and the newspaper, and it is duly printed as an authentic piece of writing.

Because of the danger of charges of fabrication, newspapers are extremely careful over handling ghosted material, using it only out of necessity and with the 'author's' full co-operation and signature on copy. The feature writers used for ghosting are adept at picking up nuances and shades of expression in conversation which they use in the writing, and they are expected to carry out their role sympathetically. The best are often professional freelances whose skill brings them a living out of this sort of work.

In production, ghosted material is handled with the same care as the written work of big-name outside contributors, any changes in text or cutting to length being checked back for approval.

READER PARTICIPATION

Reader participation is the stuff of features pages. Not only are readers' views aired in regular letters columns but their opinions are solicited on special subjects on which it is intended to run 'what you think' features – the Government's popularity, favourite holiday stories, the popularity of TV programmes, the state of the roads, for instance. Advice or service columns also live off letters, while competitions could not survive without the willing and hopeful participation of readers.

This promotional aspect of the paper is carried out largely by the features department, helped by the promotions department where there is one. It serves the dual purpose of providing the readers with a platform, and often a service, and also of getting the paper publicized and talked about.

Yet the editorial use of letters can be fraught with danger. Readers write in hundreds and thousands to newspapers each week with no idea of the limitations of space into which they are going to have to fit if used (less than one in twenty letters to *The Times*, for instance, gets published).

Choosing the letters to be used is the easy part – topicality and originality are looked for – but editing them to fit can court complaints of misrepresentation as perhaps fourteen to twenty paragraphs are reduced to three of four to give the point of the letter along with nine or ten other letters. The total area available for them might come to no more than a column and a half, including space for a main headline and a number of smaller ones.

Here is an example of readers' letters after editing in the *London Evening Standard*:

> Terry Wogan arouses mixed feelings amid all and sundry – flattery and abuse, but no one has yet pointed out his most irritating and frustrating habit.
>
> He never allows his guests to finish what they have to say. He is the most proficient interrupter on TV. It's so rude and so annoying, especially as his guests are usually thoroughly interesting people.
> *Jenny Rainbird, Aldershot, Hants*

> I have recently heard James Prior on *Any Questions* and in an interview and I could not help but notice the similarity in voice to that of the late Kenneth Horne of *Round the Horne* fame. How about a show with James Prior called *Round the Priory*?
> *G. Adams, Pinner, Middx*

> Mr John Telford Beasley, chairman, London Buses, says that 'in trials 99 per cent of passengers did, in fact, alight quite quickly and safely' from the Superbus.
>
> Considering how many millions of passengers travel by bus every year, and considering one per cent of one million is 10,000 his statistic is not as reassuring as he seems to think.
> *Simon Levene, Temple, EC4*

It takes a careful and sympathetic subeditor to edit letters in this succinct fashion without upsetting readers. The practice on most papers is to check with their writers where the subject is controversial or very personal. In some cases – on *The Times* for example – letters are cut only by permission of the writers and the cut version read back to them over the telephone. Letters are also checked with the writer if there is a fear that, though newsworthy, they might not be genuine or be part of some publicity stunt, a practice not unknown among newspaper letter writers.

The aim of letter editing is to reproduce the essential point the reader is making by using their own words with the minimum of alteration, even if a good deal of the letter cannot be used for space reasons. Well known in newspaper offices is the dictum of the Press Council that in a letters column 'editing should be done solely to qualify a letter for publication, and it should never be allowed to defeat or obscure the point or points the correspondent wanted to make'. In other words, subeditors should beware of taking phrases away from their context.

It is the custom to protect the writers from cranky correspondents by withholding the full address, unless the address (i.e. the place of origin) is part of the essential point of the letter.

In more general letters features, readers' views are quoted as part of a pattern of opinion justifying the aim of the feature. When they are used in advice columns, usually only the relevant part of the inquiry is published.

The bond with the reader is closer in provincial and town papers which provide a valuable community service by highlighting local problems and activities and giving depth and background to local news situations, as well as providing a forum for readers' views.

FEATURES AND THE LAW

Features pages give the office lawyer a lot more trouble than the news pages since the opinion and criticism they contain are more likely to wound than objective reporting of fact. Some of the more sensational 'confessions' and 'lid off it all' series that appear in national Sunday papers can suffer legal revision – and applications to the court for injunctions – right up to press time and have been known to die the death at the eleventh hour because of a court's intervention.

In buying such series, newspapers are aware that they are often heavily dependent on what the lawyer will allow in copy because of the danger that lurks in the law of libel and contempt of court. The editor's belief that the public has a right to know what he has uncovered, however true he knows it to be, has to be tempered by his preference not to have to go to jail.

Investigative journalism, in which matters of public concern, often involving crime, are examined in depth by teams of special reporters who might get evidence from unorthodox sources, are a particular headache to the office lawyer.

A subeditor is usually given tricky jobs like these to edit at the end of the legal vetting process, but the need for vigilance remains. A shift in a delicate situation or legal second thoughts can still result in a page being ripped apart at an inconveniently late hour and in changes and re-editing being carried out, and even pictures replaced, after the type has been set and the page is being made up. Speed is of the essence at this point in editorial production.

PROJECTION

The subeditor is more involved with the projection of the edited material on features pages than is the case with news. A main feature is 'sold' visually to the reader by means of a complex of headlines, pictures, blurb, standfirst and quotations in which the subeditor has a promotional role, not only editing the copy but persuading the reader of its importance and drawing attention to special aspects.

Thus editing ideas go beyond headline and intro and require a degree of identity with the aims of the writer and an ability to exploit display techniques. The subeditor is drawing out the mood and underlying substance of the feature and transmuting these in terms of picture, caption and significant quotation as well as headline. The editing, in short, is more visual and the subeditor is integrally involved in the design of the page.

Even in stock features occupying regular slots, such as a nature column or arts reviews, mood is an important ingredient of the headline and the subeditor has to be responsive to this.

A *blurb*, which a feature might require, is best described as a piece of self-advertisement carried by the newspaper, not necessarily on the same page as the feature, inviting readers to turn to it. Sometimes it consists only of one selling line that identifies the story (for example, confessions of the shy princess or your verdict on the great drugs debate) accompanied by a page reference; or it might have several sentences of description. The idea is to present a compelling reason, or reasons, why the reader should read the feature. Blurbs are mostly set in a bold display that will stand out from the rest of the page. This might include a picture as well as typography, with maybe the use of type reversed as a WOB or BOT. They combine the visual role of a display advertisement with that of a news bill.

A *standfirst* (see Chapter 14) is likely to be used on a feature, especially a series, which arises from special circumstances, to explain how or why the article came to be written. The aim is to justify or qualify what follows so as to enhance the reader's understanding or acceptance of it.

Both blurbs and standfirsts call for the imaginative and concise use of words so that within a small compass the reader can be given compelling reasons for reading on.

A useful exercise for students in the creative approach to features subediting would be to imagine a situation in which, for example, the effect of loneliness on the aged in a big city is being investigated. Take, as the ingredients, three separate reports, an exceptionally good mood picture worth running, say, as a deep horizontal six-column illustration, two other smaller pictures of interior shots and some statistics from the local council. From these, devise one main headline and two subsidiary ones, a standfirst and a panel of statistics, and project the material:

1 As a broadsheet display on a page with modest advertising.
2 As a double page spread in a tabloid paper with no advertising.

A separate static feature could be used on the page as a foil to the main projection.

A number of scenarios like this could set a pattern for workshop projects. These could also include devising a static layout slot for television and radio programmes, and one for a weekly 'name' columnist writing pithy comments on current news, in which witty one-line headings in, say, 14 pt, might be used.

FEATURES HEADLINES

Headline writing, as with page design, has a freer approach with features than with news. The headline must still draw attention to the contents of the text and form a focal point in the visual display, but these aims can be merged in a big feature into a general projection in which pictures play a bigger part (most news stories are fairly short and do not have pictures), and in which blurb, standfirst, display quotations and even crossheads and by-line all have a role.

While the headline might sum up the vital facts of a feature, it need not necessarily do so. For a start, many features are centred around comment or opinion arising out of facts that are already known. The headline is more likely to hinge on what the writer is trying to say or what the pictures on the page symbolize.

Take, for an example, the features spread from the *Liverpool Echo* (page 234). Here the main headline says: IT BEATS HELL OUT OF THE DOLE QUEUE. On its own, and set over the text with nothing else to guide the reader, this would be a meaningless message. Yet right next to the headline, and given even greater prominence, is a close-up picture of a boy boxer throwing a punch straight towards the reader. Under the headline is a horizontal picture of boys working out in a gym with a boxing trainer. In the context of the pictures into which they fit, the words 'beats hell' now make sense. They are part of a combined assault on the reader.

But what about 'dole queue'? Here, the white-on-black standfirst comes to the reader's aid. Boldly placed next to the headline and the intro, it explains: 'More and more Merseyside youngsters are turning to boxing as a way of fighting the despair of unemployment. Danny Buckland reports from where the fists are flying as youngsters bid to make the fight game big time'. Thus it gives both information and by-line.

The matter is clinched by the white-on-black strap-line that crosses the gutter at the top and knits the two halves of the spread together. It contains the quote: 'Boxing can be the only way out for these kids'. The projection is complete. The subsidiary headline on the second part of the spread down the page, JOEY'S LONG, HARD ROAD, has merely to introduce the character in the piece in such a way that, though about a different person, it is visually understood to be part of the same projection.

The composite headline-with-picture approach is also dominant in the features page reproduced from the *Daily Mail* (page 239). Whichever of the two came first in the planning of the main feature is immaterial. Each needs the other to deliver the writer's message. The mood of the piece is established at first sight. STARVED INTO SUBMISSION! is an unashamed demand upon the readers' sympathy. In such cases the verb becomes subservient to the message and the headline label acquires a visual potency of its own, even if in some readers its emotion might arouse contradiction.

The 'loaded' label approach permeates the rest of this *Daily Mail* page. THE PAWN THAT CAME TO LIFE; AN OFFER YOU CAN REFUSE . . . the reader is invited to dip into the text by an oblique approach to his or her curiosity. The appeal of the Comment column is for those who want to know what the *Mail* has to say about the issues of the day. The strategy of the main feature, by comparison, represents a direct combined assault upon the reader, who might be reading the paper for the first time. It is good visual features journalism, and such headlines have little connection with the factual purpose of news headlines.

The example of *The Guardian* features page carries to an even greater extreme the ascendancy of picture and standfirst. The *Guardian*'s style of layout, as we explained in Chapter 14, is asking the reader to accept a concept in which the headline is being deliberately downgraded in importance in relation to text and picture. All three headlines, THE COMIC STRIP TEASE, OPERATION OBERON and AMERICA'S ROAD SHOW are verbless labels which are, in fact, comments on the text. They tease the reader in the sense that they compel him or her to read the text to penetrate their meaning. Writing headlines of this sort can be dangerous, not because they are labels, but because they fail to establish a link with the reader. In this *Guardian* page it is the pictures and the standfirsts that 'sell' the stories. Although the overall visual effect is attractive, both pictures and standfirsts have not established contact with the headline, which performs merely the role of a typographical focal-point in the visual display.

There is nothing wrong with a label headline on a feature. It has a distinguished pedigree on static features such as the editorial opinion. THE BRITISH DISASTER on the opinion column on the left-hand side of the *Liverpool Echo* features spread, is a legitimate and telling comment on the prevalence of drugs and violence in British society, which the paper is comparing with the Ethiopian famine disaster.

In their modest way, the label headlines in the *Chichester Observer*'s feature page (page 240) perform a similar function. Yet an excessive reliance on an easy label can rob a feature of a potentially good headline and reader contact, considering the scope and variety of approaches that are possible. In fact, features headlines in terms of potential, can be usefully divided into five categories:

The emotive phrase

This is a good way of utilizing a writer's personal approach to a subject as in AMID THE COLOUR BLINDNESS, A VISION OF HOPE, which appears in *The Times* in a Bernard Levin article on the race problem in South Africa. Likewise, the *London Evening Standard*'s MUD THAT CAN

STICK TO THE INNOCENT, on a reader's letter about the publicity given to rape cases, uses emotive phrasing to raise a point of public concern.

The whimsical phrase

This goes well with the less serious sort of feature as in the *Brighton Evening Argus* WORDS OF WISDOM, which boldly ties together a well-pictured interview and look-back on the career of comedian Norman Wisdom. To show that whimsy can be serious as well, *The Times* has THE UNDER-WRITING ON THE WALL, on its financial page, in a pointed reference to problem being encountered by Lloyds underwriting syndicates. Whimsy, it will be noticed, can exist happily without verbs.

The informative phrase

This type of headline, on the other hand, is better with a verb, even though it can still manage without the active voice. HOW JOHN CLEESE BECAME AN EX-NEUROTIC tells baldly, and yet intriguingly, the substance of an interview by Lynda Lee-Potter in the *Daily Mail*. The active voice is given an airing with the informative command, TEST-DRIVE INTO A LEAD-FREE ERA on the same paper's motoring column, in which the writer's views are projected directly at the reader – a useful ploy in service features which aim to advise readers.

The decorative phrase

This is the resort of the subeditor when confronted by a feature full of useful titbits for the reader, odd gripes, and a few sotto voce asides which a regular columnist feels entitled to make, or for the seasonal holiday feature. The *Liverpool Echo*'s WE DO LIKE TO BE BESIDE THE SEASIDE always strikes a chord with the reader provided it's used only once a year, while the *London Evening Standard*'s HAS THE 'NYET' MAN LEARNED TO SAY 'DA'? is a useful escape for the subeditor on a piece of inconclusive Big Power political gossip.

The confessional

My fifth category remains the tour de force of all features headline approaches and the one that sells papers. The *News of the World* has unrivalled experience in this field and can come up in the same issue with the blockbusters MY ELECTRIC NIGHT SHIFTS WITH SORAYA and AGONY OF LOVING MY EASTENDERS CO-STAR. *The Guardian* provides the avant garde version of this genre with 'I'M ALWAYS HOPING THAT I WILL WEAR OUT BEFORE MY CLOTHES DO' and 'I CAN'T HELP FEELING THAT IF ALL THOSE PEOPLE LIKE MY PICTURES AND REGARD THEM AS FUNNY, THEY CAN'T REALLY BE ART' – which occur on the same page of the same issue.

It will be seen from the above examples that space and word count are not usually the problem with features headlines that they are with news since the page pattern is less formularized and the number of items to fit in are fewer. Even with the constraints of narrow measure which can apply with static features in regular single-column or double-column slots, there is greater freedom of wording because of the variety of headline approaches that are possible with features material.

The skilled features subeditor should find that the use of mood and colour in headlines and the freedom from hard news concepts should remove dependence on headlinese even in tight headlines, and such tired cliché words as probe, shock, horror, quiz, rap and drama, which can mar news page headlines, should not have to be resorted to.

GLOSSARY

ABC Audit Bureau of Circulations, the body that authenticates and publishes newspaper circulation figures.

Ad Advertisement

Ad dummy The blank set of pages of an edition with the shapes and positions of advertisements marked in.

Ad rule The rule or border separating editorial matter from advertisements on a page.

Add Copy added to a story already written or subedited.

Advance Printed hand-out of a speech or statement issued in advance to the press.

Advertising agency An organization that prepares and designs advertisements for clients, and buys advertising space.

Agony column A regular feature giving advice on personal problems to the mainly young; hence agony aunt.

Alts Alterations made to copy or set matter.

Angle A particular approach to a story.

Angling Writing or editing a story from a particular angle, i.e. to bring out a particular aspect of its news content.

Art Pertaining usually to design and layout of pages, the use of pictures and typography in newspaper display.

Art desk Where page layouts are drawn in detail and the pictures edited.

Art editor The person responsible for the art desk and for design of the newspaper.

Artwork Prepared material for use in newspaper display.

Ascender The part of a letter that rises above its x-height, as in h, k, l and f.

Author's marks Corrections or amendments by the writer on an edited story, either on screen or on proof.

Glossary

Back bench The control centre for a newspaper's production, where sit the night editor and other production executives.

Backgrounder A feature giving background to the news.

Back numbers Previous issues of a newspaper.

Bad break Ugly or unacceptable hyphenation of a word made to justify a line of type. *See Justify.*

Banner A headline that crosses the top of a page – also *streamer*.

Base The solid piece of metal upon which type rises in metal founts.

Bastard measure Any type setting of non-standard width based on columns.

Beard The space between a letter and the edge of the base upon which it rises in metal founts.

Beat An exclusive story or one that puts a newspaper's coverage ahead of another's.

Big quotes Quotation marks larger than the typesize they enclose, i.e. used for display effect.

Big read A long feature covering many columns – usually an instalment of a series.

Bill A newspaper poster advertising the contents of the paper at selected sites.

Black A copy or carbon of a story, an electronic duplicate of a story; also used to describe certain boldface types.

Blanket Newspaper page proof.

Bleach-out A picture overdeveloped to intensify the blacks and remove the tones – useful in producing a motif to use as a display label on a story.

Blobs Solid black discs used in front of type for display effect, or for tabulating lists.

Block A metal engraving of an illustration.

Blow-up Enlargement of a picture or type.

Blurb A piece of self-advertisement composed of type, and sometimes illustration, used to draw a reader's attention to the contents of other pages or issues to come.

Bodoni A commonly used serif type, noted for clean lines and fine serifs.

Body The space taken up by the strokes of a letter – the density of a letter.

Body matter The reading text of a newspaper.

Body type The type used for reading text.

Bold Name given to type of a thicker than average body.

Border A print rule or strip used to make panels for stories, or for display effects in layout – used in stick-on tape form in paste-up pages.

BOT Type reversed as black on tone background.

Box A story enclosed by rules on all four sides – also *panel*.

Break 1 Convenient place to break the text with a quote or crosshead; 2 The moment of happening of news.

Breaker Any device such as a quote or crosshead which breaks up the text in the page.

Break-out A secondary story run on a page with a main story, usually on a feature page.

Brevier Old name for 8 pt type.

Brief A short news story, usually one paragraph.

Bring up An editing instruction meaning use certain material earlier in a story.

Broadsheet Full size newspaper page approximately 22 in by 15 in, as opposed to tabloid, half size.

Bromide Emulsioned stiff paper on which photographic material is printed; any photographically printed material.

Bucket Rules on either side and below tying in printed matter to a picture.

Bulk Old name for bench or stone at which metal type was assembled after being set; also *random*.

Bureau The office of a news agency; in the US any newspaper office separate from the main one.

Buster Headline whose number of characters exceed the required measure.

By-line The writer's name at the beginning, or near the top, of a story.

c & lc Capital letters and lower case of type.

Cablese Abbreviated text used in copy transmitted by telegraph, i.e. to save transmission costs.

Caption Line(s) of type identifying or describing a picture.

Caps Capital letters of type.

Case Tray holding founts of metal type.

Case room Room where cases of type are kept and where matter is typeset or composed.

Caslon A traditional-style seriffed type face used for headlines.

Cast Printed plate taken from page made up of metal type.

Cast off To edit to a fixed length; (n) the edited length of a story as estimated.

Catchline   Syllable taken from a story and used on each folio, or section, along with folio number, to identify it in the typesetting system.

Centre spread   Material extending across the two centre-facing pages in a newspaper. *Spread*: any material occupying two opposite pages.

Centred   Type placed equidistant from each side of the column or columns.

Century   Much used modern seriffed type with bold strokes.

Change pages   Pages that are to be given new or revised material on an edition, or on which advertising material is being replaced.

Chapel   Union organization within a printing house or newspaper office.

Characters   The letters, figures, symbols, etc. in a type range, hence *character count*, the number of characters that can be accommodated in a given line of type.

Chase   Metal frame in which pages are made up in the hot metal system.

Circulation   The number of copies of a newspaper sold, i.e. in circulation; hence *circulation manager*, the executive in charge of distributing copies and promoting circulation, also *circulation rep* (representative).

City editor   Editor of financial page; in US the name given to the editor in charge of news-gathering in main office.

Clean up   Editing instruction to improve tone of copy.

Cliché   A well-worn, over-used phrase.

Cliffhanger   A story that still awaits its climax or sequel.

Close quotes   Punctuation marks closing quoted material.

Close up   To reduce space between words or lines.

Col   Short for column.

Cold type   Photoset type, i.e. not hot metal.

Colour   Descriptive writing.

Column   Standard vertical divisions of a newspaper page; hence column measure.

Column rule   Fine rule marking out the columns.

Columnar space   Vertical space separating one column of matter from another.

Command   A keyboarded instruction to a computer

Comp   Compositor; a printer who composes typeset material or makes up a page.

Compo   Composite artwork made up of type and half-tone.

Condensed type Type narrower than the standard founts; hence *extra condensed* and *medium condensed*.

Content Material in a newspaper.

Contents bill Bill or poster advertising a story or item in a newspaper.

Copy All material submitted for use in a newspaper; hence *copy holder*, a proof reader's assistant; *copy paper*, newsprint offcuts on which copy is written.

Copy-taker Telephone typists who take down reporters' copy on a typewriter or VDU (US telephone reporter).

Copy-taster Person who sorts and classifies incoming copy in a newspaper.

Copyright Ownership of written or printed material.

Corr Short for correspondent.

Correct To put right typesetting errors.

Correction Published item putting right errors in a story.

Correctors of the press Proof readers.

Count The number of characters in a line of type.

Coverage The attendance at, and writing up, of news events; also the total number of stories covered.

Credit Usually the photographer's or artist's name printed with an illustration; hence *credit line*.

Crop To select the image of a picture for printing by drawing lines to exclude the unwanted area.

Crosshead Line or lines of type to break the text, placed between paragraphs.

Cross reference Line of type referring to matter elsewhere in the paper.

CRT Cathode-ray tube, used as a light source to create the type image in a photosetter.

Cursive Any flowing design of type based on handwriting.

Cursor Electronic light 'pen' on VDU screen, used to manipulate text during writing and editing.

Cut To reduce a story by deleting facts or words.

Cut-off A story separated from the text above and below by type rules making it self-contained from the rest of the column; hence *cut-off rule*.

Cut-out Half-tone picture in which the background has been cut away to leave the image in outline.

Cuttings Catalogued material from newspapers cut out and stored in a cuttings library for future reference (in US clippings).

Glossary 259

Cuttings job A story based on cuttings.

Cypher A character in a type range which represents something else, i.e. ampersand (&) and £ and $ signs.

Data base The material to which a computer gives access.

Dateline Place and date of a story given at the top.

Dead Matter discarded and not to be used again.

Deadline Latest time a story can be filed, accepted or set.

Deck One unit of a headline.

Decoder A device for turning transmitted material into usable form, i.e. pictures or text.

Define To specify on a computer screen the material a command is intended to cover.

Delayed drop An intro which reserves the point of a story till later.

D-notice An official instruction to editors that a story is subject to the Official Secrets Act and therefore should not be used.

Descender The part of a letter that projects below the x-line.

Diary 1 The newsroom list of jobs for the day or week; 2 A gossip column in a newspaper.

Didot point Unit of type measurement slightly larger than the British-American point and used in Europe, except Britain and Belgium. Equal to 0.01483 of an inch.

Direct input The inputting of material into a computer by writers for the purpose of screen editing, i.e. by the use of VDUs.

Directory A list of stories of a given classification held in a computer and available to those with access.

Disaster caps Large heavy, sanserif type, used on a major (usually disaster) page one story.

Disclaimer A printed item explaining that a story printed previously has nothing to do with persons or an organization with the same or similar name as used in the story.

Display ads Advertisements in which large type or illustration predominate.

District reporter Reporter working from a base away from the main office.

Double The same story printed twice in the paper.

Double-column Across two columns.

Dress Redress or revision of a story; also *rejig*.

Drop letter An outsize initial capital letter on the intro of a story; also *drop figure*.

Drop quotes Outsize quotes used to mark off important quoted sections in a story.

Dummy Blank copy of the paper, sometimes half size, showing the position and sizes of the advertisement and the space available for editorial use; also mock-up of editorial pages as preparation for a new format.

Dupe Duplicate.

Earpieces Advertisements on either side of the masthead, or centred title-piece, of a newspaper's page one.

Edit Prepare copy for the press.

Edition An issue of the paper prepared for a specific area; hence *editionize*, to prepare such.

Editor Chief editorial executive who is responsible for the editing and contents of a newspaper.

Editorial The leading article or opinion of the paper.

Editorialize To insert, or imbue with, the newspaper's own opinion.

Editor's conference Main planning conference of a newspaper.

Egyptian A type family which has heavy 'slab' serifs.

Ellipsis Omission of letters or words in a sentence, represented by several dots.

Em Unit of type measure based on the standard 12 pt roman lower case letter 'm'; also called a *mutton* (in US a pica).

Embargo Request not to publish before a nominated time.

En Half an em – based on the standard roman lower case letter 'n'.

Etching The process of engraving on metal used for half-tone and line blocks in hot metal printing.

Ex's Short for expenses.

Exp Expanded (of type).

Execute Computer command meaning to put into effect.

Facsimile Exact reproduction of an original, as in facsimile transmission of pages from one production centre to another by electronic means.

Family All the type of any one design.

Feature Subjective articles used in newspapers, as opposed to objective news material; newspaper material containing advice, comment, opinion or assessment; sometimes any editorial content other than news.

File A reporter's own computer imput; to send or submit a story; a writer's or agency's day's output.

Files Back issues.

Filler   A short news item of one or two paragraphs.

Filmset type   Photoset type.

Fit-up   Artwork involving several elements joined together.

Flash   Urgent brief message on agency service – usually an important fact.

Flashback   A story or picture taken from a past issue.

Flatbed press   Small, mostly old-time, press that prints from a flat surface, i.e. not rotary.

Flimsy   Thin paper carbon copy of story.

Flong   Papier-maché board used to take moulds from pages made up in the metal.

Flush   Set to one side (as of type).

FOC   Father of the chapel (of union chapel); *MOC* (mother of the chapel).

Fold   Point at which the paper is folded during printing; hence *folder*, a device attached to the press which does this.

Folio   Page.

Follow-up   A story that follows up information in a previous story in order to uncover new facts.

Format   1   The shape and regular features of a newspaper; its regular typographical appearance; 2   Any pre-set instruction programmed into a phototypesetter.

Forme   The completed newspaper page or pair of pages when ready to be made into a printing plate.

Fount   All the characters in a given size of any type.

Frame   The adjustable easel at which paste-up pages are made up from photoset and photographic elements.

Freelance   Self-employed person, i.e. journalist.

Free sheets   Newspapers that rely solely on advertising income and are given free to readers.

Front office   Usually the advertising and editorial part of a newspaper office to which the public are admitted.

Fudge   Part of the front or back page of a newspaper where late news is printed from a separate cylinder 'on the run', sometimes called the 'stop press', i.e. the presses are stopped so that the late news can be fudged in.

Full out   Typeset to the full measure of a column.

Galley   Shallow long metal tray on which metal type is gathered and proofed before being taken to the page; hence *galley proof*.

Gatekeeper   Sociologist's name for the copy-taster.

Gatherers Journalists who gather and write material for a newspaper – a sociological term.

Ghost writer One who writes under another's name; one who writes on behalf of someone else.

Good pages Pages that do not have to be changed for later editions

Gothic Family of sanserif type with a great variety of available widths – medium condensed, extra condensed, square, etc.

Grant projector A device that projects a picture or type on to a screen to enable it to be measured for the purpose of page design, or for the preparation of artwork.

Graphics Usually any drawn illustrative material used in page design.

Grass hand A casual print worker.

Grot Abbreviation for Grotesque, a family of sans headline type.

Gutter The margin between two printed pages.

Hair space The thinnest space used between letters in typesetting systems.

Half lead The second most important story on a page.

Half-tone The reproduction process, consisting of dots of varying density, by which the tones of a photograph are reproduced on a page.

Handout Pre-printed material containing information supplied for the use of the press.

Hanging indent Style of typesetting in which the first line of each paragraph is set full out and the remaining lines indented on the left.

Hard copy Typewritten or handwritten copy, as opposed to copy entered into a computer.

Hard news News based on solid fact.

Head, heading Words for headline.

Header The part of a VDU screen in which commands and basic instructions are entered, and in which the computer communicates with the user.

Heavies Name sometimes given to the quality or serious national press as opposed to the popular press; newspapers that specialize in serious news.

H & J Computer term for 'hyphenated and justified', meaning that the material has been prepared on the screen in the length and sequence of equal lines in which it will be typeset.

Hold To keep copy for use later; also *set and hold*.

Hold over To keep typeset matter for later; also to *HO*.

Hood Lines of type above a picture or story and attached by rules top and side.

Glossary 263

Hook A term used in some computer systems for a queue or desk to which stories can be sent after tasting to await possible use.

Horizontal make-up Page design in which stories and headlines cross the page in several legs as opposed to being run up and down.

Hot metal The traditional printing system in which type is cast from molten metal into 'slugs' for assembly into pages.

House style Nominated spellings and usages used to produce consistency in a given newspaper or printing house.

Imprint Name and address of the printer and publisher, usually found at the bottom of the back page of a newspaper.

Indent Material set narrower than the column measures, leaving white space either at the front or at both sides.

Ink fly A fine spray that can hang in the air in rotary printing operations.

Insert Any copy inserted into a story already written or typeset.

Intro The introduction or beginning of a story.

Investigative journalism A form of reporting in which a news situation is examined in depth by a team of reporters under a project leader, i.e. as an investigation of all aspects.

Issue All copies of a day's paper and its editions.

Italic Type characters that slope from right to left.

Jack line A short line left at the top of a column (usually avoided in page make-up). Also a *widow*.

Journalese Newspaper-generated slang; shoddy, cliché-ridden language.

Justify To space out a line of type to fit a nominated width.

Kern Any part of a type character that sticks out from its main body, as with italic type, and is supported on the shoulder of the next character; hence kerning, to fit a line by allowing characters to fill part of the space of the adjoining characters.

Keyboard The panel of keys on a typewriter or VDU by which copy is entered on to paper or a screen.

Kicker A story in special type and setting that stands out from the main part of the page.

Kill To erase or throw away a story so that it cannot be used.

Label A headline without a verb.

Layout The plan of a page.

Lead (pronounced leed) The main story on a page; the page lead.

Lead (pronounced led) The space between lines of type under the hot metal system, achieved by using strips of metal, or leads, of set points width.

Leader Editorial opinion, or leading article.

Leg Any portion of text arranged in several columns on the page.

Legal kill A legal instruction not to use.

Legman A reporter who assists with gathering the facts but does not write the story; hence leg work.

Letterpress A method of printing from a raised or relief surface, as with metal stereo plates on rotary presses.

Letter-spacing Space the width of an average letter in a given type.

Lift To use, and keep in a page, matter that has appeared in a previous edition.

Light box A device consisting of a ground glass screen illuminated from below through which pictures can be viewed face downwards so that they can be cropped and scaled on the back.

Light face Type of a lighter weight or character than standard.

Lineage Computation of lines used as a basis of payment to writers; sometimes used for payment of non-staff newspaper contributors.

Line block An engraved plate, as in the hot metal printing system, which reproduces the lines of a drawing in continuous black, as opposed to the half-tone block which renders tones by means of dots of varying density.

Line-caster Any hot metal typesetting machine.

Line drawing Drawing made up of black strokes, as with a cartoon or comic strip.

Lino Short for a Linotype line-casting machine.

Literals Typographical errors.

Lithography Printing by means of ink impressed on a sheet.

Local corr A district correspondent.

Logo Name, title, recognition word, as of a regular column or section of newspaper.

Long primer Old name for 10 pt type; also l.p.

Lower case Small, as opposed to capital, letters of an alphabet.

Ludlow A range of headline type produced on a slug from hand-assembled matrices, also a specific typeface.

Machine minder Operator in charge of a press.

Make-up The act of making up a page, sometimes the page plan.

Masking Excluding part of a photograph by paper overlay to indicate area to be printed.

Master The basic type shape inside a photosetter from which printed type is generated.

Glossary

Masthead The name or title of a newspaper at the top of page one.
Mat Matrix, or mould (of type or page).
Mat box Box used to transmit page moulds, or matrices for plate-making and printing elsewhere.
Matrix The mould from which a page is cast, under the hot metal system; also the die from which type characters are cast.
Measure Width of any setting.
Medium A weight of type between light and bold, or heavy.
Memory The part of a computer that retains information fed into it; where written and edited stories are stored.
Merchandizing Information about price and place of purchase in consumer journalism features.
MF Abbreviation for more to follow.
MFL More to follow later.
Milled rule A Simplex rule or border with a serrated edge as on the edge of a coin.
Montage A number of pictures mounted together.
Mood picture (or shot) A picture in which atmosphere is more important than content.
Mop-up A story that puts together information already used in separate ways, or on separate occasions.
Morgue Old name for newspaper picture and cuttings library.
Motif Drawing or picture used to symbolize a subject, or to identify a feature or story.
Mould *See Matrix.*
MS Manuscript of any text before printing.
Mug shot Picture showing only a person's head.
Must An item that must be used, and containing *must* in its instructions.
Mutton Printers' old name for an em.
Nationals Newspapers on sale all over the country.
New lead A version of a story based on later information.
News agency An organization that collects, edits and distributes news to subscribing newspapers.
News desk The newsroom, where the collection of news is organized, and where reporters are based (in US, city desk).
Nibs News in brief.
Night editor The senior production executive of a daily paper.

Nonpl   Nonpareil, the old name for 6 pt type.

Nose   The intro or start to a story; hence to *re-nose*.

NS   Newspaper Society, an association for provincial newspaper proprietors in Britain.

Nuggets   Small items of news; separate sections of a story.

NUJ   National Union of Journalists (in Britain).

Nut   Printers' name for an en; hence *nutted*, type indented one nut, or nut each side.

Obit   Obituary item.

Off its feet   Type not standing straight in the page and therefore not proofing properly (in hot metal practice).

Offset   Printing by transferring the page image from smooth plastic printing plate to a rubber roller which then sets it off on to paper.

Off-stone   When a page is ready to be made into a printing plate.

Open quotes   Punctuation marks denoting the start of a quoted section.

Overline   A line of smaller type over a main headline; also a *strapline*.

Overmatter   Left-over printed material not used in the edition.

PA   Press Association, home national news agency in Britain.

Page facsimile transmission   Method by which completed pages are digitized and reduced to an electronic signal for transmission by wire or satellite to another printing centre for simultaneous production.

Pagination   The numbering of pages; the number of pages to work towards.

Panel   Story enclosed in rules or borders; *see Box.*

Paste-up   The method of making up pages from photoset material by attaching them to a page card.

Photoset   The name usually given to photocomposed type under computerized printing systems; hence *phototypesetter*.

Pica   12 pt type; unit of measurement based on multiples of 12 points (pica = one em).

Picture desk   Where collecting and checking of pictures is organized; hence *picture editor*.

Pie   To drop, muddle or disarrange lines of type (in hot metal).

Plate   Printing plate, of metal or plastic, derived from the page image.

Platen   Surface which holds the paper in a typewriter or printing press and presses it against an inked surface.

Point   Unit of type measurement. The British-American point is 0.01383 in, or about one seventy-second of an inch. *See also Didot point.* The size of type is measured by depth in points.

Populars   Mass circulation newspapers of popular appeal.

Print   Total number of newspapers printed of one issue; also a picture or bromide printed from a photographic negative.

Print order   The number of copies of an issue ordered to be printed.

Print-out   A copy of material in a computer printed out for reference or filing in advance of actual typesetting. Print-outs sometimes show the type as it will look when set.

Printing plate   The plate, metal or polymer, from which the page is printed.

Process engraving   The system of transferring a photographic image to a metal plate by etching with acid, as in hot metal printing production.

Processor   The part of a typesetter that actually produces the bromide print of the type; or produces the print from a photographic negative.

Projection   The display and headline treatment given to a story in the page.

Promotion   Any form of planned publicity that has a specific aim.

Proof   An inked impression taken from typeset material (of hot metal); also a photocopy of a paste-up page or of advertising material.

Proof reader   Person who reads and corrects proofs to ensure that copy has been accurately followed; hence *proof marks*, corrections marked on a proof.

Publishing room   Where the newspapers are counted, wrapped and prepared for distribution.

Puff   An item in a newspaper which publicizes something or somebody.

Pull-out   Separate section of a newspaper that can be pulled out, often with separate pagination.

Pundit   A regular newspaper columnist who dispenses opinion.

Qualities   Serious, as opposed to popular, newspapers.

Queue   A collection, or directory, of stories held in a computer – features queue, newsroom queue, etc.

Quire   Unit of freshly printed, ordered newspapers, usually twenty-six copies.

Quotes   Raised punctuation marks to indicate quoted speech.

Qwerty   Standard keyboard layout as on a typewriter or VDU, based on the first five characters of the top bank of letter keys.

Ragged (left or right)   Copy set justified on one side only, sometimes used in captions.

Random   Bench or table in the composing room where type is assembled and checked before being taken to the pages.

Range   The number and variety of characters available in a type.

Rate card List of newspaper advertising charges based on specific sizes and placings.

Reader A proof reader.

Reader participation Editorial material or items which involve contributions by readers, such as readers' letters, competitions and articles based on invited opinions.

Readership The total number of people who read a newspaper – not the number of copies in circulation. The estimated number of readers per copies of magazines and newspapers can vary considerably.

Redress *See Rejig*.

Reel Spindle holding a roll of newsprint; sometimes, a roll of newsprint; hence *reel room*, where rolls of newsprint are stacked for use.

Register The outline of printed matter as it appears on the paper; important in colour printing where the main colours are printed separately on to the picture image.

Rejig The revision of a story in the light of later information, or a change of position in the paper, often between editions.

Release The date or time handout material becomes available for use.

Re-nose To put a new intro on to a story, using different material or a different angle.

Re-plate To replace a printing plate to allow a later version of a page on to the press.

Reporter Person who gathers and writes up news.

Retainer Periodic payment made to retain someone's services, as with local correspondents; *see Stringers*.

Retouching Improving the quality of a photographic print by the use of a brush or pen.

Reuters British based international news agency.

Revamp General change given to a story or page in the light of a reconsidered approach.

Reverse Type printed white on a black or tone background; can be done in a photosetter as *reverse video*.

Reverse indent *See Hanging indent*.

Revise To check and correct, or improve, edited material.

Rewrite To turn a story into new words rather than to edit on copy.

RO Run on (on typed copy).

Roman The standard face of a type.

ROP Run of press. For instance spot colour is printed during run of press rather than as a separate or additional process.

Rota picture A news picture obtained under the rota system, in which limited coverage of an event is allowed on a shared basis.

Rotary press Traditional press in which newspapers are printed by the letterpress method from curved metal or polymer relief plates.

Rough Outline sketch of page layout.

Rule A printed border of varying width.

Run Length of time taken to print an issue of a newspaper.

Running story A story that develops and continues over a long period.

Run on To carry on printing without changing plates for an edition.

Rush Second most urgent classification of news agency material after *flash*; hence *rushfull*, a full version based on rushes.

Saddle A metal mount used for attaching polymer plates to a rotary press to achieve correct printing height.

Sale or return Newspapers sold subject to a fixed payment for unsold copies carried.

Sans Sanserif, types without tails, or serifs, at the end of the letter strokes.

Satellite printing Printing at subsidiary production centres by the use of page facsimile transmission.

Scaling Method of calculating the depth of a picture to be used.

Scalpel Used to lift, cut up and place material in paste-up pages.

Schedule List of reporting or feature jobs to be covered for use in an issue of a newspaper.

Scheme To plan and draw a page; also a *page layout*.

Scoop Exclusive story.

Screamer Exclamation mark.

Screen The density of dots in half-tone reproduction of photographs.

Screen The part of a VDU on which stories held in the computer are projected for reading or editing; hence *screen subbing*, subbing by electronic means by use of a cursor.

Scroll (up or down) To display material on to a VDU screen so that it can be read in sequence.

Seal Standard words, often in spot colour, at the top of a page indicating the edition; also a logo.

Section A separately folded part of a newspaper; hence sectional newspapers.

Send A command to transfer material in a computer to another queue or desk, or to the typesetter.

Separation The separate elements of a colour picture by which the colour is transferred to the page.

Sequence The order in which a story is presented (in subbing).

Series Range of typesizes, or of types.

Serif Type characterized by strokes that have little tails, or serifs.

Service column An advice, or consumer, column.

Set and hold Put into type for use later.

Set flush To set full to the margin.

Set forme The last forme (of page or pair of pages) to go to press.

Set solid To set without line spacing.

Setting format Setting of a nominated size, width and spacing that is programmed into the computer, i.e. where such setting is regularly used.

Shorts Short items of edited matter, usually of one, two or three paragraphs.

Sidebar Story placed alongside a main story to which it relates.

Side-head A headline or cross-head set flush left, or indented left.

Sign-off The name of the writer at the end of a story.

Situationer A story giving background to a situation.

Sizing *See Scaling*.

Slab-serif Type with heavy square-ended serifs.

Slip To change a page between editions; hence *slip edition*.

Slug Line of type cast on a line-casting machine.

Slug-line Catchline (in US).

Snap Piece of information in advance of full details in news agency story.

Spike Home for unwanted stories. Computers also have an electronic *spike* to which stories can be sent.

Spill To run down and fill space (of type).

Splash The main page one story.

Split screen The use of a terminal to display two stories at once.

Spot colour Non-processed colour applied to the page during run of press.

Spread A main story that crosses two adjoining pages.

Squares Black or open, a species of type ornament used to mark off sections of text.

Stand-first An explanation in special type set above the intro of a story, i.e. it stands first.

Stand-up drop   An initial letter in large type that stands above the line of the text at the start of a story.

Star   Type ornament; hence *star-line*, a line of stars.

Star-burst   Headline or slogan enclosed in star shaped outline, used in blurbs and advertising.

Start-up   When the presses begin to print.

Stet   Proof reader's mark means 'as it stands'.

Stick   Small metal tray used for hand-composing metal type.

Stock bills   Newspaper display bills on fixed subject such as 'today's TV', 'latest scores', etc.

Stone   Bench where pages are made up under hot metal system; hence *stone sub*, the journalist who supervises this work; *stone-hand*, the printer who works on the stone.

Stop press   Late news printed from a separate cylinder on to the page while on the press, or afterwards.

Strap-line   Headline in small type that goes above the main headline; also overline.

Streamer   Headline that crosses the top of the page, also a *banner* headline.

Stringer   A local correspondent.

Subeditor   Person who checks and edits material for a newspaper to fit set space, and writes the headline (US deskman).

Subhead   Secondary headline.

Subst head   Headline in place of another.

Syndication   The means by which a newspaper's material is offered for a fee for use in other publications or countries.

Tabloid   Half size (newspaper).

Tag-line   An explanatory line or acknowledgement under the bottom line of a headline.

Take   A piece of copy, part of a sequence.

Tear-out   A picture printed with a simulated torn edge, usually a flash back of a printed picture, or a part of a document; also *rag-out*.

Telephoto lens   Camera lens that magnifies an image telescopically.

Teleprinter   Machine that prints text received by telegraphic signal.

Terminal   The part of a video display terminal on which electronically-generated text is displayed and monitored.

Textsize   A broadsheet, or full-size newspaper page.

Tie-in   A story that is connected with one alongside.

Tie-on A story that is connected with the story above.

Tip-off Information from an inside source.

Top A top of the page story; mostly any story that merits a good headline and of more than three paragraphs in length.

Trim To cut a story a little.

Turn head A head covering a story that has been continued from another page (US a *jump head*).

Typebook Catalogue of types held.

Typechart A tabulated list giving character counts for given types.

Underscore To carry a line or rule under type.

Update To work in later information.

Visualize To plan and work out how a page or display will look.

VDU Video display unit, a device with a screen and keyboard used to display and enter text into a computer; also VDT, video display terminal.

Web-offset A system of printing in which the inked page image is transferred from a smooth printing plate on to a rubber roller and then offset on to paper, as opposed to being printed directly on to paper by relief impression.

Weight The thickness of a type.

Widow A short line left at the top of a column of reading type; also *Jackline*.

Wing in To set a headline within the top rule of a panel or box, leaving a piece of rule showing on either side.

Wire A means of transmitting copy by electronic signal which requires a receiver or decoder; hence *wire room*, where such copy is transmitted or received.

Word processor Electronic system by which text can be keyboarded into a computer, stored, edited, amended and finally printed (i.e. processed) when required.

Work station A special video display terminal used at a distance from the computer, with access and facilities to enable work to be done away from the main production centre.

WOB White on black type.

WOT White on tone type.

x-height The mean height of letters in a type range, exclusive of ascenders or descenders.

INDEX

Abbreviations, 55, 135, 143
Accuracy, 77–81, 242–3
Advertising:
　content, 33–4
　production, 200–3
　revenue, 34
　space, 23, 34
Agence France Presse, 17
Alternative spellings, 132
Alternative words, 162–9
American words, 133
Anglicized words, 133
Apostrophe, 116
Associated Press, 17, 31
Authors' and Printers' Dictionary, 132

Bad copy, 179
Bans, 134
Baptist Handbook, 80
Bartholomew's Gazetteer of Great Britain, 181
BBC Year Book, 80
Bias, 229
Birmingham Post, 110
Blurbs, 29, 249
Brackets, 114
Burke's Peerage, 80
By-lines, 54, 232, 233, 250

Camera:
　equipment, 61
　telephoto, 61–2
Capa, Robert, 61
Capital letters, 133
caption writing, 182–5
Cartier-Bresson, Henri, 61, 68
Caslon type, 47, 241
Casting off, 94–6
Century Schoolbook type, 47, 48, 238
Catholic Directory, 80
Chichester Observer, 241, 251

Chief subeditor, 35
Circulation revenue, 34
Circumlocutions, 121
Civil Service List, 81
Clichés, 122–4
Cold type, 1, 198
Colloquialism, 222
Colon, 113
Columnists, 233, 235–6, 249
Commas, 113
Community service, 248
Composing:
　make-up, 202–9
　paste-up cutting, 207
　room, 199
　spacing, 208
　stone subbing, 206–7
Compositor, 203–9
Computer revolution, 1, 4–5, 7
Confessions, 248
Contempt of court, 173
Contents:
　balance, 34
　bills, 185–6, 229
　features, 226–7
Copytaster, 14, 19–23
Copytasting:
　electronic, 30
　flow, 226
　method, 30
Creative writing, 227
Crime, 248
Crockford's Clerical Dictionary, 80
Cropping (of pictures), 67
Crossheads, 53–4, 232, 238
Cuttings library, 79
Cyphers, 135

Daily Mail, 148, 220, 238–9, 250–2
Daily Mirror, 12, 21, 223
Daily Mirror Style, 123, 223

274   *Index*

Daily Star, 12
Daily Telegraph, 21, 27, 29, 66, 215, 245
Dashes, 114
Date lines, 54
Debrett, 80
Decentralized printing, 11
Delayed drop intros, 221
Direct input, 5, 13, 187
Dod's Parliamentary Companion, 80
Drop letters, 53, 238

Editionizing, 14–15, 180–2
Editor, 19, 35, 228
Editor's conference, 33
Editorial manning, 35
Electronic copytasting, 30–2
Electronic editing, 13, 96–9, 180, 187, 198
Ellipsis, 115
Ethics, 73
Evans, Harold, 62, 147
Evening Argus, Brighton, 24–5, 54, 66, 218, 252
Everyman's Dictionary of Quotations, 81
Excelsior type, 205
Exchange telegraph, 31
Exclamation mark, 116

Facetiousness, 229
Familiarity (in page planning), 34, 253
Fashion writing, 229
Features:
 agencies, 227
 control, 230
 copy sources, 227
 display, 233
 freelances, 227
 investigative, 229
 logos, 232
 outside contributors, 227
 page design, 231
 page make-up, 230–1
 pictures, 231
 planning, 226–30
 policy, 228–9
 production, 229
 series layouts, 233
 slots, 233
 staff writers, 227
 text, 232
 topicality, 229
 typography, 231
 work flow, 228
 workshop, 233, 249
 writers, 226
Features editing, 242–53
 and the law, 248
 language, 243
 projection, 248
 style, 244
Features headlines, 233, 250–3
Fibre optic cables, 33
Financial columns, 219
Financial Times, 12, 21–2, 39, 40–1, 44–5, 140, 216
Follow-ups, 22
Foreign Office List, 81
Foreign words, 120
Format (of paper), 33, 35
Formatting, 201
Free Church Directory, 80
Freelance reporters, 16
Fudge, 14
Full page composition, 12, 65, 202–3

Galley proofs, 206
Garst and Bernstein, 162
Geography, 93
Ghost writing, 245–6
Gill Sans type, 47
Graphics, 74–6
Grot no. 9 type, 47
Group correspondents, 227

H & J-ing, 7, 205
Handouts, 17
Hart's Rules For Compositors and Readers, 132, 206
Headline writing, 137–69
 abbreviations, 143
 adjectives, 140
 alternative words, 162–9
 composition, 146–8
 content, 144–6
 creative moment 160–2
 direct approach, 146
 feature headlines, 233
 label headlines, 149
 location, 146
 nouns as adjectives, 140
 numbers, 144
 oblique approach, 147

Index 275

Headline writing – *continued*
  omission of words, 139
  personal touch, 145
  punctuation, 141
  special words, 139
  split headlines, 149
  subject, 138
  symbols, 140
  taste, 144
  things to avoid, 150
  thoughts, 148–51
  time, 145
  turn heads, 150
  verb, 139
  vital facts, 144
  words, 138, 141, 162–9
Headline typography, 152–62
  arrangement, 154
  character counts, 155–8
  spacing, 158–60, 205–9
  type style, 155
Headlinese, 253
Heswall and Neston News and Advertiser, 181
Hodgson, F. W., *Modern Newspaper Practice*, 227
Hot metal, 1–4, 8–10, 206, 208
House style, 131
Hoylake News and Advertiser, 181
Hutt, Alan, 154
Hyphen, 118

International Who's Who, 81
Intros, 52–3, 83–8
Investigative journalism, 248
Ionic type, 205

Jane's All the World's Aircraft, 80
Jane's Fighting Ships, 80
Jenkins, Simon, 245
Journalese, 225

Layout artists, 35, 52
Leader (editorial column), 232, 236, 241
Leader writing, 229
Lineage, 17
Linecaster, computer-driven, 8
Linotype machine, 2–3, 199–200
Literals, 3
Liverpool Echo, 66, 217, 233, 235, 250–2

Local correspondents, 16
Logos, 232
London Evening Standard, 92, 110, 222, 247, 251–2

Magazine sections, 226
Make-up, 202–9
Malton Gazette & Herald, 181
Marking up (copy), 97
McCullin, Don, 61
Measure (of typesetting), 51
Mergenthaler, Ottmar, 3
Misused words, 128
Multiple copy sources, 178

Name style, 134
Newcastle Evening Chronicle, 186
News:
  agencies, 17
  balance, 34–5
  creation, 14
  editor, 18
  measurement, 21
  stereotypes, 82
News of the World, 12, 79, 148, 252
Newsroom, 18
  electronic, 18
Nouns, 108
Numbers, 134

Official Secrets Act, 175
Order (in subbing), 81–2
Oxford Companion to English Literature, 80
Oxford Companion to Music, 80
Oxford Dictionary of Quotations, 81
Oxford English Dictionary, 132

Page design:
  elements, 46
  features pages, 231, 234–41
  focal points, 44
  principles, 37–46
Page facsimile transmission, 11
Page planning:
  essential points, 35
  order, 45
  purpose, 33
Panels, 54
Paragraphs, 111
Participles, 102
Penguin Dictionary of Quotations, 81

Penguin English Dictionary, 163
Photographers:
  briefing, 61
  equipment, 61
  freelances, 59
  staff, 58
  transmitters, 61
Photosetting, 7–8, 49–51, 55, 200–1, 203, 205–6
Pictures:
  agencies, 59
  at work, 66
  balance, 63
  bleach-outs, 66
  choice, 64
  collected, 60
  colour, 57
  composition, 63
  conditioning by, 63
  control, 56, 62–3
  cropping, 67, 235–8
  cut-outs, 65
  design function, 64–6
  digitized storage, 62
  editing, 58, 61–2, 67–74
  ethics, 73
  features use, 233
  general, 56–76
  handout, 60
  library, 60
  montage, 65
  reproduction, 65
  retouching, 71
  reversing, 72
  rota, 60
  scaling, 70
  sequence, 65
  sizing, 70
  sources, 58–61
  taste, 73
  tone, 63
  uses, 64–6
Pickering Gazette & Herald, 181
Platemaking, 2, 11, 205, 209–12
Playbill type, 48
Points system, 49
Polymer plates, 10
Pravda, 12
Prepositions, 109
Press Association, 17, 31, 179
Printing systems, 1–13
  colour, 72–3

Privilege:
  absolute, 176
  qualified, 176–7
Production, electronic aids, 12–13
Programmable keys, 205
Pronouns, 103
Proof reading, 206
Public and Preparatory Schools Year Book, 80
Punctuation, 112–18

Qualifiers, 107
Question mark, 117
Quotation mark, 117
Quoted speech, 89

Radio and TV Who's Who, 80
Random (of print), 200–2
Readers:
  letters, 227
  participation, 227, 246–8
Readership market, 37–44
Redresses, 182, 188
Rejigs, 182, 188
Retouching, 71
Reuters, 17, 31
Reviews, 238
Revise sub, 37, 180
Rinehart, William D., 203
Rockwell bold type, 48, 238
Roget's Thesaurus of English Words and Phrases, 163
Rota pictures, 60
Rotary presses, 8–10
Royal Family, 21
Running stories, 187–98

Sanserif type, 46–9
Scaling (pictures), 70
Scheming pages, 37
Screen page make-up, 12, 65, 202–3
Semi-colon, 113
Sentence structure:
  general, 101–11
  length 109–11
Serialization, 226, 233
Serif type, 46–9
Service columns, 227
Shorter Oxford Dictionary, 150
Side-bars, 53
Sign-offs, 232–3
Situationers, 15

Soft sell technique, 220–2
South London Press, 215–16
Special vocabularies, 213–20
Split infinitives, 106
Sport, 217–19
Staff reporters, 16
Stand-firsts, 232, 249
Stereotypes, 124
Stresses, 118
Stone, 3, 199
Stone-subbing, 206–7
Style, 227, 243–6
Subediting:
  accuracy, 77–81
  angling, 177
  background, 93
  bad copy, 179
  basic techniques, 77–99
  casting off, 94–6
  delayed drop intros, 87–8
  editionizing, 180–2
  electronic aids, 4–8, 180, 198
  features, 242–53
  geography, 93
  intros, 83–8
  law, 170–7
  making up, 97, 199–212
  multiple copy sources, 178–9
  order, 81–3
  quoted speech, 89–93
  rejigs, 182, 188
  revising, 180
  rewrites, 177–9
  running stories, 187–98
  sequence, 88–93
  stone subbing, 206
  time, 93
Subeditors, 35
Syndication, 227
Synonyms, 122, 163

Tabloid style, 223–4
Technical words, 120
Tenses, 104
The Air Force List, 80
The Army List, 80
The Directory of Directors, 81
The Guardian, 2, 21, 66, 91, 151–3, 236–8
The Navy List, 80

The Observer, 34
The Sun, 12, 24, 27–8, 54, 66, 140, 223–4
The Sunday Times, 34, 42–4, 75, 210–12, 245
The Stock Exchange Year Book, 81
The Times, 21–2, 148, 213, 216, 218, 246–7, 251–2
Time sequence (in subbing), 93
Times New Roman type, 47
Trade names, 136
Type:
  balance, 52
  character, 45
  faces, 46–7
  features, 231
  instructions, 51
  measure, 51
  setting, 51
  sizes, 49
  style, 135
  use, 48–9
  variants, 48–9, 52–5
Typography, 46–51

Underscores, 55
United Press International (UPI), 17

Vacher's Parliamentary Companion, 80
VDUs, 4, 12, 31, 39
  portable, 19
Verbs, 105
Vogue words, 124–31

Wall Street Journal, 12, 152
Waterhouse, Keith, 123, 223
Web-offset printing, 8–11, 209–12
Western Mail, Cardiff, 26–7, 66, 213, 216
Whitaker's Almanack, 81
Who's Who, 80
Who's Who in America, 81
Who's Who in Music, 81
Who's Who in the Theatre, 81
Word processors, 19
Word use, 119–36

X-height (of type), 50

Yorkshire Post, 217